AF704283

PROPANE CONVERSIONS

BY JAY STORER

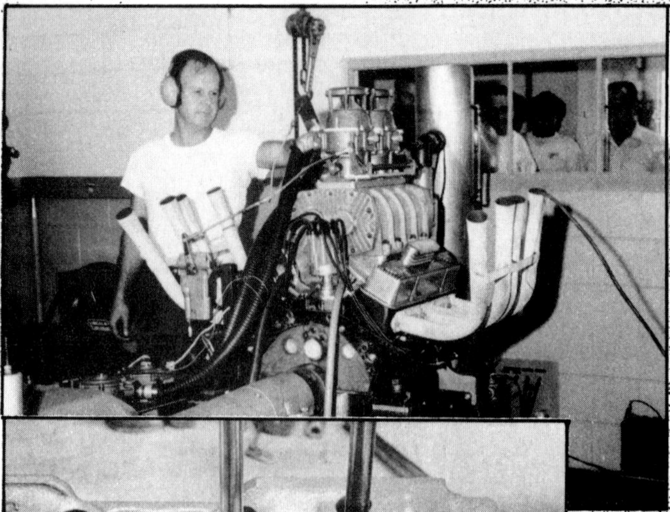

The information contained in this publication is sold without any warranty as to its useability or performance. All the risk for its use is entirely assumed by the purchaser/user. In all cases, original manufacturer's recommendations, procedures, and instructions supersede and take precedence over descriptions herein. Specific component design and mechanical procedures—and the qualifications of individual readers—are beyond the control of the publisher, therefore the publisher disclaims all liability incurred in connection with the use of information contained in this publication.

PROPANE FUEL CONVERSIONS

EDITED BY LARRY SCHREIB
PRODUCTION BY LARRY ATHERTON

ISBN 0-931472-12-1

Copyright © 1986 by S-A Design Books, 515 West Lambert, Bldg. E, Brea, CA, 92621-3991. All rights reserved. All material in this manual is the copyright property of S-A Design Publishing Company. It is unlawful to resell, redistribute, or copy this information in any way without the expressed written permission of the publisher. Printed in U.S.A.

S-A DESIGN BOOKS, 515 WEST LAMBERT, BLDG. E, BREA, CA 92621-3991

CONTENTS

CHAPTER 1
INTRODUCTION

In addition to domestic resources and pipeline supplies from Mexico and Canada, large quantities of LP gas are imported, usually shipped via huge tankers built especially for hauling LPG. Future needs will require additional facilities for unloading and storing imported LP gasses.

What would be your reaction to a newly announced "super fuel" that promised to provide internal-combustion engines with greater efficiency, decreased emissions, longer engine life, and reduced maintenance? After your initial skepticism, your first questions might well be, "What's this new fuel going to cost, where can I get it, and how soon can I tank up?"

You may be shocked to learn that this super-fuel costs less than gasoline, is available all over the U.S. (in thousands of locations), and has been available as a motor fuel for over 30 years! The fuel we're talking about is not a new discovery or the result of a laboratory breakthrough. It's simply *propane gas*, otherwise known as *LPG* (for *liquefied petroleum gas*). It's abundant, clean, efficient, and supported by an industry with proven technology for its implementation as a fuel for engines ranging from monstrous industrial powerplants displacing 10,000 cubic inches down to those of lawnmower size. It shows tremendous promise as an alternative fuel; and no matter if your main interest is finding fuel for your motorhome, avoiding gasoline lines, cleaning up the atmosphere, coping with inflation, or searching for the ultimate racing fuel, it may very well be the fuel in your future.

WHAT *IS* PROPANE FUEL?

Thanks to an award-winning animated commercial produced by the Chevron Oil Company, even the most unscientific-minded of us now realize that animal and vegetable matter from bygone ages has been trapped underground for eons and transformed into various carbon forms that modern man finds useful. In addition to the valuable discoveries of diamonds, coal, and oil formed from carbon, man's drilling and digging has brought him other useful hydrocarbons, not the least of which is natural gas.

Today natural gas is available in almost all areas of the world and has been used as a fuel for perhaps several thousand years. It's believed that the famous Greek Oracle at Delphi may have used natural gas for the continuously burning flame. Early Chinese had gas wells, some as deep as 2000 feet, from which they transported the gas through bamboo pipes and burned it for heat and illumination. By 615 B.C., it is known that Japan had similar wells. These early drillers had discovered the flammable gas when drilling for underground pools of natural salt brine, from which they could distill salt (a precious commodity of the time). This early Oriental use of natural gas was abandoned, for some unknown reason, about 1700 years before Western Man had ever learned of its existence.

As found in nature, natural gas is a mixture of several gasses but is composed largely of methane (CH_4). This composition often includes hydrogen-sulfide (H_2SO_4), which we commonly identify with the smell of rotten eggs. Natural gas that contains hydrogen-sulfide is aptly called "sour gas," while that without it is called "sweet gas." A typical sweet gas

Floating somewhere in the Java Sea, the 460ft Ardjuina Sakti is a floating LPG refrigeration and storage terminal. Owned by Arco, the ship—shown here loading a standard LPG transport ship—handles 375,000bbls and was named one of the ten outstanding engineering achievements of 1977 by the National Society of Professional Engineers. This type of storage can effectively supplant present land-based storage facilities.

contains 82% methane, 4% ethane, 3% propane, 3% butane, 2% hexane, 5% heptane, and miscellaneous amounts of other heavier hydrocarbons.

Propane—the portion of natural gas that we are concerned with—is the third member of the alkane (or methane, or paraffin) group of hydrocarbons, which is at the lighter end of the hydrocarbon spectrum. Its natural state is a colorless, odorless, non-toxic vapor ($CH_3CH_2CH_3$ or C_3H_8), and the commercial propane available today is extracted from natural gas—one of the *refinery gasses* that are produced as byproducts of the crude-oil refining process. We may be accustomed to thinking of both natural gas and propane as having distinctive odors, but this is due to odorants added by the gas distributors as a safety feature (to make leak detection easier).

PROPANE AS A MOTOR FUEL

Hydrocarbon gasses have been used as motor fuels since a Swiss named Isaac DeRivas used "illuminating gas" to build the first internal-combustion-engine-powered vehicle in 1806. In 1862, the French inventor Lenoir installed his new internal-combustion-engine design in a carriage and powered it with gas. Five years later, Nicholas Otto, the German forever remembered as the man who perfected the four-cycle internal-combustion engine (we now call it the Otto-cycle engine), exhibited his engine running on illuminating gas at the Paris Exposition. In the 1920s, the elephantine Zeppelins of Germany were powered by Blau Gas, which the Germans had cracked from crude oil.

Today, the use of natural gas and propane fuels is widespread throughout Europe, where they have been used for the past 30 years in commercial vehicles and taxi-cabs. And throughout the world, countries that have large reserves of gas and low reserves of oil—including Canada, Australia and New Zealand—are rapidly converting their vehicles and fleets to gaseous fuel so they can become less dependent on foreign oil for their future energy needs.

In Japan, where energy is a daily crisis and air pollution is high (smog problems are even worse in Tokyo than in Los Angeles), major efforts have been made to convert city-use vehicles, such as delivery trucks and taxis, to low-pollution propane fuel. Almost 300,000 vehicles in Japan are already on gaseous-fuel diets, and some 30,000 are being converted each year. Both Toyota and Datsun perform propane conversions right on their factory assembly lines (sorry, these conversions are not exported to the U.S.).

Since the 1930s, natural gas and propane have been used in the U.S. as motor fuels (chiefly in stationary engines, tractors, and forklifts); but with carburetor improvements developed in the 1950s, these fuels are now used in a multitude of applications.

Natural gas (CNG) has a multitude of uses and is available through a tremendous network of underground and above-ground pipes that supply the stoves, water heaters, furnaces, air-conditioners, clothes dryers, and hot tubs of millions of Americans (not to mention the vast amount of natural gas that is used to fuel thousands of stationary commercial powerplants that produce electricity or pump water, liquids, and gasses through pipeline networks). In

The most readily available source of supply is a standard LPG dealership that sells bulk and retail. This is the storage tank at a typical gas dealership. Motor fuel is generally cheaper at these dealerships than at roadside gasoline stations or RV/camper supply centers.

these days of energy consciousness, we tend to think of the United States as running on oil, *without realizing the extent to which natural gas also affects our daily lives.*

However, there's another gaseous fuel more important than we realize: the propane gas produced from natural gas and crude oil. During the "cracking" process, about 3% of a barrel of crude oil becomes propane. The propane produced by this process answers the energy needs of over 18 million customers, which ranks propane as the fourth most important source of national energy.

There are many areas of the United States where pipeline natural gas is not available as a fuel for home or industry, simply because it is uneconomical to pipe natural gas to areas that have a very low population density; so propane provides an excellent alternative in these areas. In their natural state, gaseous fuels are quite voluminous; and to transport them by means other than through pipelines (i.e., special tanks or pressure vessels), they have to be compressed into liquid form. Since it takes tremendous pressure to force natural gas into a compressed or liquid state (for efficient transportation), tanks built to hold liquefied natural gas (LNG) must safely withstand a pressure of 2250psi! Since natural gas *boils*—or turns from liquid to gas at -258°F (-161°C)—LNG storage tanks must also be built from special metals that can take the extreme cold without becoming brittle. This requires expensive alloys of magnesium/aluminum or nickel/steel.

By contrast, propane boils/freezes at -43.7°F (-42°C); and the gauge pressure of the liquid (the pressure of the vapor above the liquid—these tanks cannot be filled to 100% capacity with liquid) is 175.8psi. So tanks for liquid propane or LPG can be made of standard steels. Because they are pressure vessels, they have to be round; but they don't have to be long and skinny like conventional high-pressure tanks. (CNG tanks are similar in configuration to the oxygen bottles used with gas welding torches.)

Obviously, propane is much easier to liquefy (LPG), store, and transport than natural gas; and some 60 million Americans depend on it in one way or another. The distribution of approximately 33.8 million metric tons annually (a metric ton equals approximately 500 gallons) takes an industry involving more than 86,000 people and a capital investment of $7.5 billion. Currently, propane is stored in 250 primary storage facilities (with capacities totaling over 7 billion gallons) and is distributed by a network of 8000 bulk points and 25,000 retail outlets. It is distributed

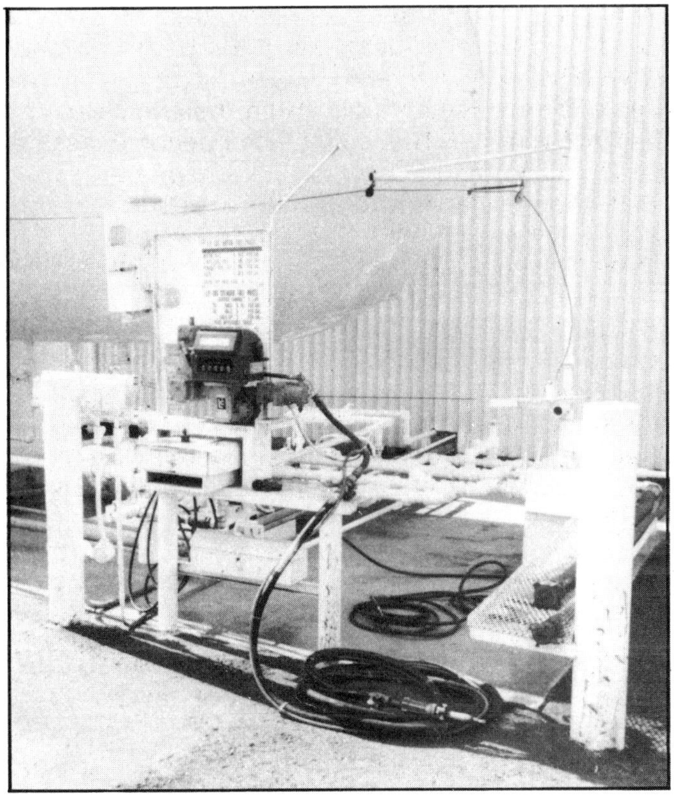

This is the meter and pump used to dispense LPG at a dealership. Hose fittings are generally available for any kind of tank, from a camper bottle to an LPG truck.

In rural areas not served by conventional natural-gas pipelines, home heating and cooking needs can be provided by onsite bulk-storage propane tanks. These tanks are refilled periodically by tanker trucks from a local LPG dealership.

Gaseous-fuel carburetion can be fitted to virtually any engine and application, ranging from huge stationary powerplants (this Waukesha V-16 develops 2000 horsepower at 1200rpm on natural gas) to something as small as a two-cylinder engine for a portable generator.

through 232,500 miles of pipelines, transported cross-country via 25,000 trucks and 22,000 railroad cars, and carried over water by more than 370 tankers and barges.

Because of the special requirements of the transportation tanks, natural gas isn't very practical as a motor fuel; and there are other limitations as well. While propane can be found at outlets all across the country (in urban areas and along highways, it is dispensed as a motor fuel or as a fuel for refrigerated truck trailers; and in rural areas, it is available at dispensing stations for filling farm and home tanks), natural gas is comparatively rare. The local gas and power company is usually the only source; and while they may use it as a motor fuel for their own trucks, they don't often sell it to private parties. In rare cases, they may make it available to a few fleet customers; but to utilize natural gas in any type of fleet operation, the company must own a huge compressor and special storage tanks in order to compress and store the natural gas (in gaseous form) obtained from the pipelines. Even though the fuel is cheaper than propane, these facilities are expensive; it's only feasible when there are existing pipelines to supply the natural gas.

For an individual user—even if a supply of natural gas is available—there are reasons why propane is a better choice as a motor fuel. The heat value of a gaseous fuel is measured in BTU's (British Thermal Units) per cubic foot of gas. Since natural gas is a mix-

ture of several gasses, it's heat value varies, depending on the source and what percentages of other gasses are mixed in with the methane. Natural gas may vary from a value of 1228 BTU/cubic feet (typical of Pennsylvania sources) to only 742 BTU/cubic feet (common in Texas fields), but it commonly averages about 1030 to 1100 BTU/cubic feet.

By comparison, propane is a relatively pure fuel (though some butane is also found in LPG) with a heat value of 2558 BTU/cubic feet—twice that of even the best natural gas. However, when you buy liquefied gasses for fuel, you don't purchase them by the cubic foot (as you do when they come through a pipeline into your house in a gaseous state). Interestingly, when compared by weight (as liquids), natural gas comes out ahead: a pound of LNG contains 23,891 BTUs, and a pound of propane contains 21,670 BTUs. On the other hand, liquefied gasses sold as motor fuels are sold by the *gallon*; and since propane weighs 4.23 lb/gallon, as compared to only 2.5 lb/gallon for LNG, the BTUs/gallon are 91,044 for propane compared to 59,727 for natural gas. In general, this makes propane a more practical and economical motor fuel.

PROPANE ADVANTAGES

Gasoline is currently our No. 1 motor fuel, and propane is our No. 3 most popular fuel. (Diesel fuel is No. 2; and all of the other "alternatives," such as gasohol, straight alcohol, hydrogen, electricity, wood, steam, etc., hardly combine to make a drop in the proverbial bucket.) So if we're to discuss propane as an alternative motor fuel, we have to compare it to

Despite what the chemistry books tell us, many engines apparently *gain* power when converted to propane fuel. This Chrysler 426cid Hemi was built by Don Bass and raced with outstanding success. The basically stock Hemi (except for dual propane mixers) developed 561 horsepower and produced consistent, trouble-free runs at the dragstrip.

Bill Chisholm may have the only early Corvette running on alternative fuel. This is a fine example of a well-planned conversion. The engine features hardened valve seats, a precision valve job, a specialty intake manifold, and a high-energy ignition.

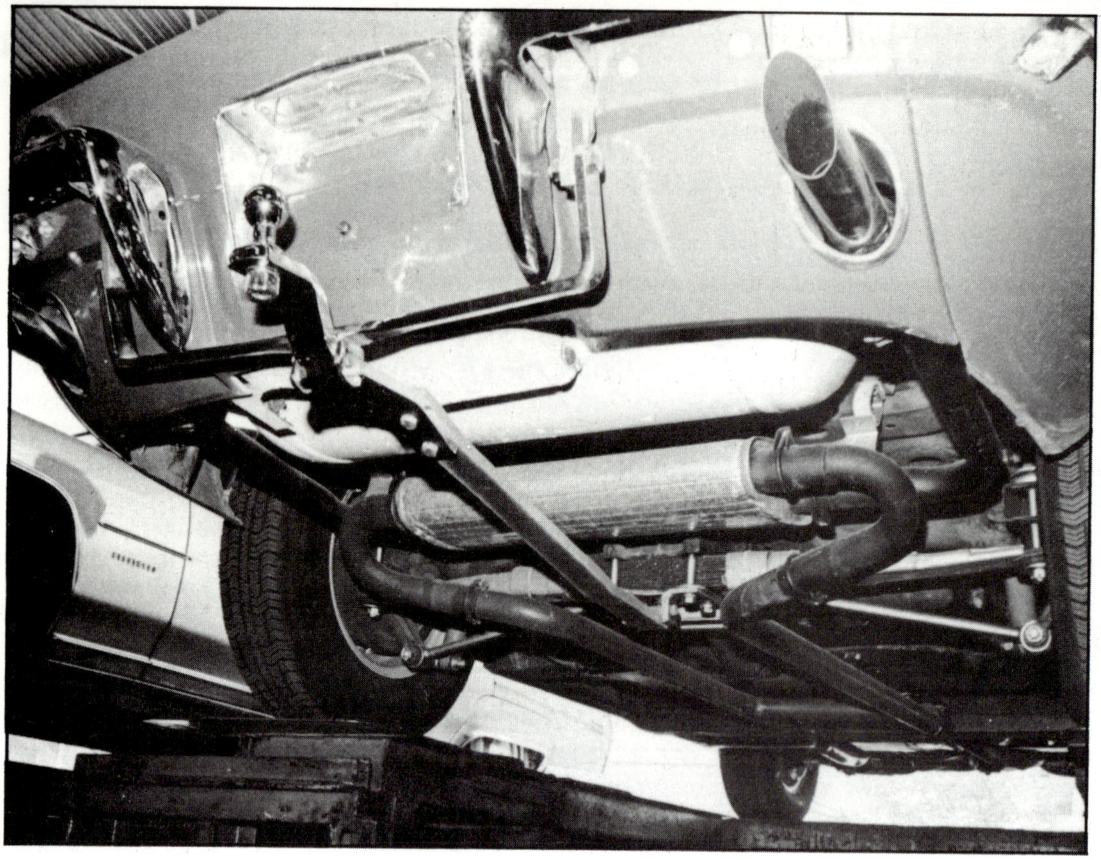

A custom-built propane tank sits under the rear deck, but you would hardly notice it from normal viewing angles. This tank gives the slick 'Vette a range of about 600 miles.

To keep the LP system inconspicuous, the tank was designed to locate the filler under the stock Corvette fuel cap. The outage valve is visible to the right of the license plate.

gasoline. The following text encompasses most of the major advantages and disadvantages of propane compared to gasoline, all of which deserve additional discussion.

One of the major advantages of any gaseous fuel is that it *mixes* so well with air. Internal-combustion engines must induct oxygen along with a suitable fuel before combustion can occur. If the fuel is a liquid, like gasoline or diesel fuel, it has to be "vaporized" and mixed thoroughly with the inducted air (which contains the required oxygen). For various physical reasons, this vaporization is somewhat difficult to achieve under all operating conditions. And although carburetion has come a long way since Nicholas Otto, none of the carburetion devices dreamed up so far (including fuel injection and electronic engine controls) can match a gaseous fuel for complete fuel and air "mixing" under all engine conditions.

It takes heat—lots of heat—to vaporize a liquid, but it has to be just the right amount. Gasoline will vaporize at around 400°F; and most modern smog engines are operated with coolant temperatures over 200°F to increase fuel vaporization (coolant or exhaust heat is usually ducted through the intake manifold to increase vaporization), reduce exhaust emissions, and gain maximum fuel economy. Gasoline engines run more efficiently at even higher temperatures (experiments have been run on engines operating at 300°F with excellent efficiency); but current cooling-system and exhaust-valve technologies are not ready to cope with these temperatures for extended periods of operation.

It is also important to remember that during short "around-town" trips (the most common operating conditions for most automobiles), the typical gasoline engine often does not reach a sufficient temperature for efficient vaporization. And, of course, during a *cold start*, economy and emissions are heavily affected—adversely—by the lack of proper fuel vaporization. This is one reason that efficiency is usually improved during longer highway driving. Under these conditions, the engine has a chance to get really warm; and the intake manifold has enough hot water, hot exhaust, and hot oil splashing through and under it to do a more efficient job of vaporizing the fuel.

Propane, on the other hand, becomes a vapor at any temperature above -43.7°F (when it is no longer held in a liquid state by pressure). So it provides 100% vapor to blend with the inducted air in a *homogenized* mixture even on cold starts, short-trip driving, or the dead of winter. *This thorough homogenization of the fuel/air mixture is the key to the advantage that propane provides as a motor fuel; and this advantage results in lower emissions, longer engine life, longer spark-plug and oil life, increased engine smoothness, no carburetion problems, and much more.*

EMISSIONS AND PROPANE

To see how propane affects emissions, it is relatively simple to operate a conventional internal-combustion, spark-ignition engine (a "pure and simple" engine with no electronic wizardry and no catalytic converters or other smog "band-aids") on a dynamometer (a device for studying engine operation under controlled conditions) with gasoline fuel and then convert it to propane fuel and take a second set of readings. Such tests have shown that an engine, on propane, will be as much as *10* times cleaner in percentages of carbon monoxide (%CO) and that total emissions may be as little as 50% of the levels emitted by a gasoline engine. (An engine on natural gas, by the way, would have total emissions of about 25% of the gasoline engine.)

The propane engine will also be cleaner in oxides of nitrogen (NOX) because there is a lower heat value in the fuel than there is in gasoline; and NOX emissions, as a rule, go up or down in direct relation with the temperature of combustion. The emissions of hydrocarbons (HC) may be improved on propane as well, but HC emissions are more a function of engine design and condition than of the type of fuel being used. An engine that has the proper spark and valve timing, a suitable compression ratio, an efficient combustion-chamber design, and a tight ring-cylinderwall seal will be low in HC on both fuels, with the propane-fueled engine generally being as good or better than the gasoline version. Conversely, an engine with excessive HC emissions on gasoline may be just as inefficient with propane fuel.

Cold starts? Winter operation? These conditions present no problem for a propane engine, either in emissions or driveability. Laboratory testing is not often conducted in extremely cold conditions—though millions of cars and trucks in the Midwest and Northeast are routinely operated during the most adverse winter weather imaginable. But many propane users can testify that propane engines do not exhibit the choking, stumbling, extended high-idle speed, and the black smoke spewing from the tailpipe (because of excessively rich mixtures) demonstrated by conventional gasoline engines. In this situation, a propane engine would really have an advantage, resulting in lower average emissions under all wea-

Gaseous fuel is widely recognized as a clean-air fuel; carbon-monoxide production is very low and combustion emissions are less photochemically reactive than the exhaust byproducts of gasoline or diesel engines.

Low emissions, quiet operation and a superb safety record make propane an ideal fuel for many specialized applications; including warehouse forklifts and public transportation, like this light tram.

ther conditions.

The emissions advantages of gaseous fuels have been documented so thoroughly that the Environmental Protection Agency (EPA) has declared that propane is a "clean-air fuel." Since 1975, when the catalytic converter and unleaded gasoline became the norms for U.S. cars, emissions of gasoline engines have been declining as the manufacturers were forced to comply with the ever-tightening restrictions mandated in the federal Clean Air Act of 1970. Because of the very low emissions of modern cars with electronic engine controls, exhaust oxygen-level sensors, and three-way catalytic converters (converters designed to clean up the exhaust in all three areas of pollution, %CO, HC, and NOX), the gasoline engine now compares more favorably with the propane engine—at least in the area of emissions.

But there are different *kinds* of hydrocarbons emitted by engines on these very different fuels, and

This elegant Maserati sports/luxury car has been fitted with a custom LPG fuel system that produces outstanding performance and complete reliability.

There are several foreign exotics that cannot be imported because they exceed strict U.S. emissions standards. However, most imports can be converted to propane, and they will then meet virtually all current emission requirements or qualify for exemption. The remote-fill of this propane-fueled '68 Ferrari fits into the stock gas-filler location.

Federal tests do not take this into account, just as they do not take into account the cold-weather emissions. In the early 1970s, the Automobile Club of Southern California engaged in a series of extensive emissions and engine-life tests on 18 vehicles in their fleet converted to propane fuel and dual-fuel (where the engine can be operated on either gasoline or propane). Among their findings was the startling fact that the hydrocarbons emitted by a propane engine (remember that propane is at the lighter end of the hydrocarbon scale) "have a much lower reactivity than those produced as a product of gasoline combustion; i.e., it takes longer to convert LPG hydrocarbons to photochemical smog. Because of the longer reaction time required, less movement of air can clear

The Weber four-barrels of this V-12 Ferrari have been replaced with a trio of Impco CA125 mixers. This converted exotic exhibits exceptionally smooth idle characteristics, instant throttle response and a 12% increase in horsepower!

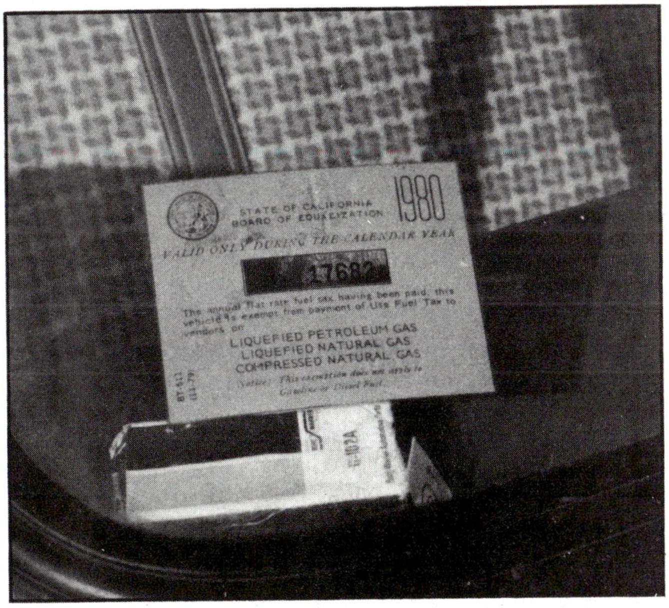

In some states the tax advantage for propane-powered vehicles takes the form of a window decal. The specific details vary from state to state—the example shown here is California—but generally the user pays road tax on just the first 600 gallons. Fuel purchased in excess of 600 gallons (per year) is essentially free of road tax.

the area of pollutants before they can react to sunlight."

California—specifically, the California Air Resources Board (CARB)—has always been a leader in the exercise of strict controls over new-car vehicle emissions. Their standards have consistently been more stringent than federal emissions requirements (so much so that the manufacturers have to include additional smog equipment on all models destined for California); yet the CARB has consistently viewed propane conversions favorably, recognizing them as clean-air equipment. In fact, at one time, the state encouraged propane conversions by allowing owners to deduct 50% of the conversion cost on state income-tax returns. Unfortunately, this is no longer the case; but there are still some inducements that indicate that the CARB believes in propane as a clean alternative to gasoline. When any owner switches a pre-1978 car from gasoline to straight propane, he can remove almost all of the original smog equipment from the engine; and if a 1978 or later engine is converted, the vapor-recovery system can be removed.

California also has a motor-fuel tax break for propane engines. After a system approved by the CARB is installed, the operator of a propane engine can purchase a sticker from the State Equalization Board. When the operator stops for a refill and displays the sticker, the fuel dealer will not charge State road-tax fee on the fuel. In the past, the fee for the sticker has been equivalent to the road tax on 600 gallons of fuel. Therefore, when the owner uses more than 600 gallons of fuel in a year, the additional fuel is essentially tax-free! If, for instance, the car or truck in question gets 15mpg on propane and the vehicle is operated for about 15,000 miles per year—resulting in a yearly consumption of 1000 gallons—the operator has saved the equivalent of the tax fee on 400 gallons of fuel. A number of states now offer similar tax stickers.

One final aspect of the emission picture is the way the engine is set up and tuned. Earlier, we compared a propane-converted engine to a similar engine designed to be operated on gasoline. This means that it was designed to build up heat in the intake tract for fuel vaporization—heat that is not only useless in a propane conversion but actually harmful to engine efficiency. However, if an engine has the benefit of a few changes that *optimize* it for running on LPG, it can perform even better in all aspects of emissions and efficiency. We'll get into the why's and how-to's of this in a later chapter; but suffice it to say that all of the factory emissions equipment of the last 10 years has been designed, with gasoline in mind, to come as close as possible to achieving an "ideal" stoichiometric air-gasoline ratio of 14.7:1. But propane operates very efficiently at an fuel-air ratio as lean as 18.0:1. As a result, when a late-model engine is converted to propane and forced to retain all of the smog equipment designed for gasoline, these devices don't necessarily perform as they should.

For instance, the typical catalytic converter is designed to reduce exhaust hydrocarbons (unreacted fuel residue); but to function efficiently, it must reach an operating temperature of 1150-1500°F. However, when the engine is fueled with propane, the converter temperature will be lower and it will not operate efficiently.

The latest cars equipped with electronic engine controls, oxygen sensors in the exhaust, and three-way catalytic converters could prove to be very efficient and clean—on propane fuel—if they could be equipped with a suitable program for the computer controls that was designed for propane mixtures instead of gasoline. One propane proponent has even suggested that a car so equipped could conceivably emit tailpipe exhaust that is cleaner than the polluted air typically entering the air filter! This statement was made, perhaps, with tongue in cheek; nonetheless, there is no question that a modern propane-fueled engine is much, much cleaner than its gasoline-fueled counterpart.

EXTENDED ENGINE LIFE

A lot of factors make up the overall operating costs of a motor vehicle: fuel costs, taxes, repairs, lubrication, tires, insurance, etc. If you're the type of individual who can afford to (and wants to) purchase a new vehicle every two or three years, then the cost of major engine repairs probably never figures into your personal automotive budget. For the rest of us who like to hang onto a good vehicle for several years, a complete engine rebuild may become an additional expense. If you spread the cost of a carburetor overhaul, valve job, or even a complete engine overhaul (usually after 80,000 to 100,000 miles) over the total mileage driven, it's easy to see that such repairs notably increase the cost-per-mile of operation. And if you're operating a vehicle fleet, with anywhere from 20 to 1000 vehicles (most of which are inevitably driven until they virtually drop dead), you can see that frequency of repairs becomes an extremely important factor in the cost of doing

Engine durability? On propane, this 390 Ford truck has traveled over 745,000 miles on the same shortblock!

Many government agencies across the country have taken advantage of LPG fuel for police cars, city fleets, parking patrols, school buses, and refuse-recovery vehicles. Longer vehicle life, less repairs and ever-shrinking budgets are compelling arguments for these conversions.

business. Not only is it expensive to maintain a crew of mechanics and an inventory of repair parts, but think of the cost of the down-time when one or more of the vehicles is in the shop for major repairs.

It is important, therefore, for some operators to think about overall engine life; and because of the dry, gaseous, light-hydrocarbon nature of propane, *engine life can usually be extended to as much as three times normal by using propane fuels*! What makes a gasoline engine wear out? Basically there are two factors: 1) the "wet" nature of gasoline and 2) the amount of carbon deposits that accumulate in the engine. Given an adequate supply of clean oil, a typical engine should be expected to last for at least 100,000 miles (if it isn't abused by the operator); but few engines seem to last that long.

The average gasoline engine doesn't always get an abundant supply of clean oil; and from the moment you fire it up, gasoline is helping to wear out engine lubricants. Over a period of time, this affects all of the moving, friction-generating parts. A good piston ring-cylinderwall seal is essential to the efficiency of an engine; yet every time a gasoline engine is cold-started, every time the throttle is pumped before starting it, every time the choke is set or accelerator pump is activated, an overly-rich gasoline mixture is supplied to the cylinderwalls. This raw gasoline dissolves the oil from the cylinderwalls and allows the piston rings to scrape against dry metal. In addition, excess gasoline seeps past the rings and mixes with the crankcase oil. This dilutes the oil and greatly reduces its lubricating properties.

In the propane engine, there is no dilution of the crankcase oil, no washing down of the cylinderwalls, no overly-rich mixtures under a cold-start (or any other condition, including high-altitude operations). Propane carburetion equipment reacts to atmospheric pressure in a way that gasoline carburetors do not and compensates automatically to lean out the mixture at higher altitudes when the air is less dense.

The other life-prolonging advantage of propane fuel is the almost total lack of carbon residue deposited inside the engine. This keeps the oil cleaner and, more importantly, doesn't provide the abrasive grit that causes premature internal wear. Hard carbon makes an excellent grinding and polishing compound; but until you disassemble an engine for a rebuild, you don't often realize how much of it is constantly being swirled throughout a gasoline engine with the lubricating oil.

What is the effective life of a piston and rings in a gasoline engine? Maybe 100,000 miles under the best conditions? Talk to the fleet mechanics of the *Los Angeles Times* newspaper trucks—all operated on propane; they'll show you pistons that have gone 250,000 miles and still look ready to re-ring and install again. Ask the propane delivery-truck driver in Las Vegas (where constant high temperatures are deadly on engines and cooling systems) who is proud of his truck having served 600,000 miles, or the cab driver who has maintained his own vehicle and gone 3/4 million miles without opening up the short-block.

Talk to propane dealers, people who do propane conversions, and fleet operators who have had millions of trouble-free miles from gaseous fuels, and you will gain some first-hand insight on the value of propane. After years of religious and frequent oil changes at short intervals on gasoline engines, ask them how often they change their engine oil on propane. It is not uncommon to hear something like, "Oh, eventually the oil does get darker; but chemical analysis of the oil always shows that it has life left in it. We usually change the oil when we lose our nerve!"

And the clean-burning characteristics of propane also extend spark-plug life to two or three times normal. At today's prices this can amount to considerable savings for the fleet operator.

While we're still discussing operating costs and repairs, think about what gasoline carburetion equipment costs over the long run. It will usually cost more than $150 to replace a carburetor on any modern gasoline engine. It costs almost that much to have it completely rebuilt, and some foreign "economy cars" have carburetors that list for over $300! While some gasoline carburetors may run trouble-free for 50,000 to 60,000 or more miles, most need some work at every tune-up interval. The main jets can become clogged; needle-and-seat assemblies wear

Motorists who face costly engine rebuilds and fleet mechanics who continually repair heavy-duty engines will appreciate the long engine life provided by LPG fuel. This piston is straight from a truck in the *Los Angeles Times* newspaper fleet. Even after 250,000 miles, it looks good enough for new rings and reinstallation!

When you're bouncing across 1000 miles of desert at flat-out speeds the simplicity and reliability of a propane fuel system can be a big advantage. These propane-fueled Ford pickups finished first in their respective classes in the 10th annual Baja 1000 race. The 351-pickup is driven by Frank Vessels and the Courier is driven by Dick Landfield.

out or malfunction; accelerator pumps break down; floats, chokes and other parts get out of adjustment; and many carburetors need "boiling out" and a rebuild kit every 30,000 or 40,000 miles.

None of these carburetion repairs are common on propane engines (at least on straight propane installations; dual-fuel conversions may still experience common gasoline-related problems). Good propane carburetors seem to run almost forever without any tinkering, and *there isn't a lot that can go wrong with them.* In the event that a diaphragm finally gives up, inexpensive repair kits are available that anyone with a little instruction can install himself. Another plus is the fact that a propane system does not require a mechanical fuel pump. Because the liquefied fuel is under 160 to 190psi pressure in the tank (we'll see later how this pressure is reduced before delivery to the propane carburetor or mixer), there's no need for a pump. Have you checked the price of a new fuel pump for your gas-burner? If not, you may be in for a real shock.

Even without requiring repairs, gasoline carbure-

The high octane rating and rugged, simple carburetion make propane an ideal fuel for racing; especially off-road racing, where contamination of liquid fuels is a constant and difficult problem.

Propane has been used widely in rural communities and farms for more than 30 years; but this is a fairly recent application—fueling the high-horsepower monsters used in competition tractor pulls.

tion systems can develop annoying idiosyncrasies that are not found in a propane system, such as the loss of power and fuel economy when running at higher altitudes (unless the carburetor is re-calibrated to compensate for the reduced air density). Because the propane is under pressure, traveling through the desert on a hot day will never sideline you with "vapor lock"; nor will you ever have to add expensive gasoline additives in freezing weather. Hard starting, flooding, and "flat spots" become things of the past with propane.

What really amazes first-time drivers is the absolute smoothness of the engine. Shakey and vibrating inline fours and sixes run with V8-smoothness; and during acceleration, there are no dead areas or hesitations. Excellent "driveability" is one of propane's hidden advantages. Fleet operators who maintain vehicles that are largely driven in stop-and-start conditions with prolonged periods of idling (such as local delivery vehicles, trash-collection trucks, taxis, etc.) find that the spark plugs do not "load up" with carbon deposits and that driveability is every bit as good as in steady-state, long-distance trips.

Another important—although not money-saving—advantage of propane over gasoline is certain to spark an interest in those who seek extra engine performance. *Propane has a relatively high octane rating.* Gaseous fuels, like natural gas and propane, are at a lighter end of the hydrocarbon scale than gasoline; and the fewer number of carbon atoms means that they aren't as volatile as gasoline (they're tougher to light) and they burn slower. Slow-burning fuels might not sound advantageous to the uninitiated, but this is precisely what gives a fuel its octane rating. Low-octane fuels, like unleaded gasoline or regular gasoline, burn quickly in the combustion chamber; this characteristic leads to detonation or "pinging" when the reacting mixture doesn't burn at the proper rate to apply pressure smoothly against the head of the piston. Slower-burning fuels produce an even push on the piston head and promote a more complete burning of the mixture.

When clean and pure, propane has an octane rating (Research) of 125; but most LPG mixtures contain some butane (octane rating of 91). Thus, what you are likely to buy may be "only" 105 octane. Even so, this is equal to or better than aviation or racing gasoline—and considerably higher than any of the premium pump gasolines available today. Propane's high octane rating makes it perhaps the ideal fuel for high-performance applications, particularly with supercharged or turbocharged engines. (With modern pump gasolines, it is not practical to take full advantage of a turbocharger without adding expensive octane boosters to avoid detonation.) Even in normally-aspirated passenger cars, pickups, and RV's, the engine will live a lot longer when it isn't "rattling and banging" the crank and rods on every mountain grade.

DISADVANTAGES OF PROPANE FUEL

Blessings always seem to be mixed with shortcomings; and as good as propane is as a motor fuel, it isn't going to be satisfactory for everyone. Inconvenience will certainly eliminate many of those who are interested only in avoiding gas lines and saving a little bit on the cost of fuel. In some areas, propane may be hard to find; and in *any* area its going to be harder to find than gasoline or even diesel fuel. Because there isn't a propane dealer on every corner, you have to plan your fuel needs and head for the nearest dealer well before the fuel gauge reads completely empty. The fuel is under pressure in the tank; so if you run the vehicle dry somewhere on the side of the road, you can't just run to the station and come back with a Dixie cup or a 10-gallon Stetson full of fuel. (This inconvenience is one of the reasons fleet operator generally maintain their own in-house refueling capability.)

Nonetheless, those who appreciate and desire the advantages of propane learn to put up with some inconvenience; and most difficulties can be readily avoided with a little conscientious preplanning. Throughout the country, many major propane distributors publish booklets that list their local dealers; so unless you're planning a day-and-night, non-stop driving binge, interstate and cross-country trips aren't a problem. You'll find propane for sale virtually everywhere, often advertised right alongside the major highways. It's not a fuel for the hurried traveler, because you may have to travel a few miles off the interstate to get to the nearest dealer; but this can be an advantage for the adventurous, bringing you into contact with people and places you would never have otherwise discovered. However, there are few "all-night" propane stations, so you will have to plan your

A typical inter-city delivery truck is an ideal candidate for propane conversion. LPG tanks can be readily fitted to the same location as the gasoline or diesel tanks.

last fill-up (if you do much driving at night) to get you to the propane dealer before normal closing hours. Propane dealers are generally open from 8:00 to 5:00 on weekdays and half a day on Saturday; but you will find that most dealers are closed on Sundays. (Some gasoline stations carry propane for RV's and maintain longer hours.)

The cost of the initial conversion to propane may deter some; but most potential LPG customers are those who plan to keep a vehicle long enough to make the conversion pay off—not just in lower fuel costs but in maintenance, too. The equipment to convert a car to straight propane may cost in the neighborhood of $500 to $600, and a complete installation with labor included may run $1000 to $1500. The pressure-vessel tank is the most expensive part of the equipment, costing $200 to $400 alone; but sometimes propane dealers or installers may be able to offer you a used tank at considerable savings.

Be aware, too, that the equipment is not always something you can walk into a store and purchase. There may be a waiting list ahead of you. After the "gas crisis" of 1973 and during past rounds of "the oil-shortage game," the demand for propane conversion equipment completely outstripped the supply. At one time, the carburetion and tank manufacturers were as much as 9 months behind on orders, though this has finally improved with recently increased production. If there's another gasoline crunch with long frustrating lines, the demand may again outstrip production. And if you do make the conversion, it may be a good idea to install a lock on the hood; during past gasoline shortages, unscrupulous people stole vehicles just to get the propane equipment, which could be sold (at the time) for outrageously inflated prices.

Two other disadvantages, both concerning your engine, are the possibility of "valve recession" and a reduction in both horsepower and fuel economy; but don't make any hasty conclusions about either of these factors, yet.

Like unleaded gasoline, propane does not contain lead or other compounds to lubricate the interface between the valve head and the valve seat. In an unmodified engine, this can mean that the valves will wear on the seats and eventually *recede* into the head. In the worst case, this will require new seats to be installed in the heads; in milder cases, a new valve job will clean up the recession. An engine unmodified for propane (or unleaded gasoline) may run for 20,000 to 80,000 miles before needing a valve job to compensate for the recession. It's a result of many factors besides the fuel alone (mainly the operating temperature of the combustion chamber and the valve material). In addition, the problem of valve recession on propane can be radically delayed and even eliminated without major expense, as we'll see in a following chapter on building an engine for propane.

The loss of mileage and power stems from the mixed blessing of the low-hydrocarbon nature of the fuel. Propane has less BTUs per gallon than gasoline; and looking at the two fuels strictly by the book, there should be a 10% to 20% loss in energy when switching to propane. (Think of an engine as burning fuel and air by *weight*; remember that propane weighs 4.24 lb/gallon, while gasoline weighs approximately 6.0 lb/gallon.) Fortunately for us, engines don't respond to fuels exactly according to chemical theories alone; and efficiency isn't strictly a matter of BTUs. Since propane burns so much better than gasoline (because it blends readily with the inducted air to form a homogeneous mixture) and since it distributes more evenly to the different cylinders without falling out of suspension and puddling along the intake tract, the power and mileage may not drop as much as the BTUs might lead one to expect.

Some modern engines operate so badly on gasoline (with all the efficiency-robbing smog equipment) that performance is actually improved when they are converted to propane. More importantly, we've been talking about comparing an engine *unmodified* for propane fuel—in which case, efficiency will suffer because most engines are designed for gasoline. A few modifications to the engine to improve it for propane (such as keeping the unnecessary heat out of the intake tract, running the engine cooler, and raising the compression ratio to take advantage of the high octane rating) will easily bring the propane engine to the same level of power and fuel economy—*and perhaps even better in the power department*. And adding a turbocharger along with propane will almost always regain all of the power and substantially more, as we shall soon see; but in any case, you can be certain that those belching diesels won't be able to run off and leave you in a cloud of smoke!

PROPANE EQUIPMENT

In order to fuel an engine with propane, four components are required: a tank, a filter/shutoff device, a vaporizer/regulator, and a mixer or carburetor. These pieces are unique to propane systems, and none are interchangeable with any comparable gasoline equipment. Additional pieces are required for a dual-fuel installation (an engine that is designed to be fueled with either propane or gasoline), but these requirements will be discussed in a later chapter.

THE TANK

The onboard fuel tank will generally be the most expensive component required for a propane conversion, and the user's specific fuel needs should be considered carefully before selecting a tank. Obviously, since there are few propane refueling stations available, it is more convenient to carry a larger tank and thereby increase the maximum vehicle range between refills.

A propane tank is a pressure vessel that must be designed and built somewhat "round" or tubular in

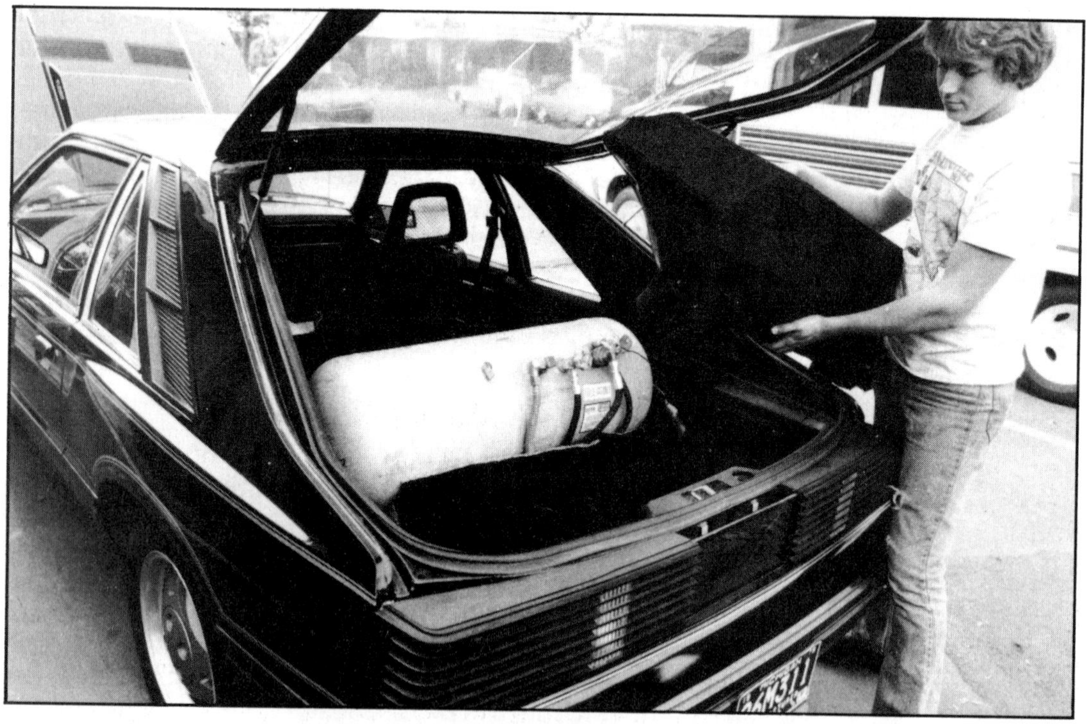

These two Chevrolet pickups represent the old and the new of propane conversions. Pickup trucks are among the most popular subjects for conversion, and tanks are available in a wide variety of sizes to fit almost any truck bed.

A passenger-car conversion is sometimes more difficult because the tank must be externally mounted or positioned in a location that is completely sealed off from the passenger compartment. In some cases, like this 1980 Capri, there is not a sealed trunk compartment, so a metal cover (in this case covered with carpet) must be fabricated.

general shape to safely contain the gas pressure. There is, however, a relatively wide variety of tanks available; so with careful research, it is usually possible to find a container of adequate length and diameter to carry a suitably large amount of fuel in the available space.

In most installations, the most difficult job—such as it is—is to determine the location and proper mounting arrangement for the fuel tank. Aside from large fleet trucks, which often have large tanks mounted under the chassis (in about the same locations as typical diesel tanks), most trucks can usually be fitted with a simple bed-mounted tank. Passenger-car installations can generally utilize a trunk-mounted tank.

In any case, it is important to carefully measure the available space and study the catalogs from various suppliers to determine the most suitable tank length and diameter for your specific installation. And when measuring the available space for the tank, it is important to consider how the tank will be lowered into the final mounting position. A tape measure will slink into a lot of places where a bulky propane tank won't fit!

It is particularly easy to convert a pickup truck to

In some cases, the LPG tank can be located under the vehicle in place of the stock tank, as on this Dodge van.

propane because the tank can be mounted easily at the front of the pickup bed, and most tank manufacturers make containers that are designed specifically to be installed in the beds of popular pickups. Pickup beds are usually deep enough to accept a quite large tank that will not protrude above the level of the bed sides (which will allow for the installation of a bed tarp). On fleetside trucks, it is also important to remember that the tank mount may have to accommodate the wheelwells, which often protrude into the bed in the area where the tank can be most conveniently located.

In passenger-car installations, the tank can be mounted in the trunk or under the rear frame, where the original gasoline tank was located. In instances where space is severely restricted, multi-cylinder tanks can be used (though these are more expensive than single tanks). For example, two or three small-diameter tanks (which are interconnected with welded steel tubes) can be fitted in the same space as the typical passenger-car gasoline tank without reducing ground clearance.

It is also possible to mount a single large tank in the trunk compartment; but in such a case, the trunk must be completely sealed off from the car interior. This is an important and essential safety measure that will prevent a propane spill from entering the passenger compartment. And just as important, it will reduce the likelihood that a fire in the passenger compartment will endanger the trunk-mounted tank. Of course, this means that a completely sealed *metal* firewall must be installed between the passenger compartment and the trunk.

THE FILTER/FUEL-LOCK

Once the tank is located and properly mounted, the next major component in the system is a high-pressure hose that carries liquid propane from the tank to a filter/fuel-lock. This dual-purpose device performs two important functions: as the name implies, it 1) filters particulate impurities from the fuel and 2) shuts off the supply of fuel when the engine is not running.

When the liquid propane enters the bottom of the filter/fuel-lock, it passes through a pad of filtering material. This filter will catch any pieces of rust or debris that may wind up in the tank or fuel lines. In a well-maintained system, this filter will not accumulate much debris (and it only needs to be inspected occasionally); but it is particularly important in a newly-installed system because it will catch miscellaneous

A unique, functional and space-saving tank mount is shown in this pickup. Two long cylinders are mounted along the sides of the bed. A fiberglass cover fits over the entire bed, and there's still plenty of room in the bed for 4x8 plywood, etc.

Where more than one tank is used, electric lock-offs (such as those used in dual-fuel installations) can be used so that a dashboard switch will supply the engine from either or both tanks.

tidbits that inevitably creep into new fuel lines, etc.

The second stage of this device serves an important safety function. Some fuel-locks are vacuum-operated and some are electrically-operated; but in either case, the fuel-lock shuts off the supply of liquid propane when the engine is turned off. The vacuum-type senses manifold vacuum when the engine is running and releases liquid propane to the vaporizer; the electric type releases fuel when the ignition key is turned on. Some experts consider the vacuum-type safer because it does not rely on an electrical contact switch (which may malfunction); and in the event of a mishap that causes the engine to stall, a vacuum switch will cut off the fuel flow immediately (with an electric fuel-lock the fuel flow will not be terminated until the ignition key is switched off).

A vacuum-operated *filter/fuel-lock* is a simple device. Liquid fuel from the tank enters at the bottom and goes through a filter pad. When the vacuum line (not shown) senses engine vacuum, the diaphragm (upper chamber) lifts and allows liquid to flow to the converter (where it is converted to a vapor).

THE VAPORIZER/REGULATOR

The next major component in the typical propane fuel system is the vaporizer/regulator. It consists of two chambers, one above the other. Liquid propane that is released from the tank by the filter/fuel-lock passes through the upper chamber, and coolant from the engine passes through the bottom chamber. Heat radiated by the coolant raises the temperature of the liquid propane; it is then converted to a vapor state, and a pressure regulator reduces the subsequent vapor pressure (in two stages) to slightly less than atmospheric.

The vaporizer/regulator is necessary because of the inherent cooling potential (latent heat of vaporization) of liquid propane. Propane absorbs a lot of heat when it is depressurized and converted from a liquid to a vapor; and if the relatively warm engine coolant was not circulated through the vaporizer, the system would freeze or "ice up" as the propane is vaporized.

THE CARBURETOR (MIXER)

When the engine is operating, a pressure drop in the manifold (vacuum) activates the pressure regulator, permitting vaporized propane to travel to the final major constituent of the system—the carburetor. A propane carburetor is a simple yet precise metering device designed to induce the proper mix of fuel vapor and air to the intake manifold. As is evident in the accompanying illustration, a typical propane carburetor is very simple (as compared to a complicated gasoline carburetor); but it will function reliably and almost indefinitely with little or no maintenance.

The best carburetor designs operate on an air-valve principle, which utilize a constant pressure drop to draw fuel into the carburetor from the regulator. These designs operate on such a comparatively low vacuum drop—typically, a vacuum drop of six inches (water)—that just cranking the engine with the starter pulls the air valve against the spring, holding it closed. This makes engine startup almost instantaneous. The first startup of the day may take 5 to 10 seconds of cranking; but thereafter, when the vaporizer is already warmed up, starting is extremely easy.

SAFETY

For the past 30 years, propane has been used as a motor fuel in this country; and when viewed objectively, the safety record of the fuel has been exemplary. At one time or another, every conceivable type of accident has occurred; and in virtually every in-

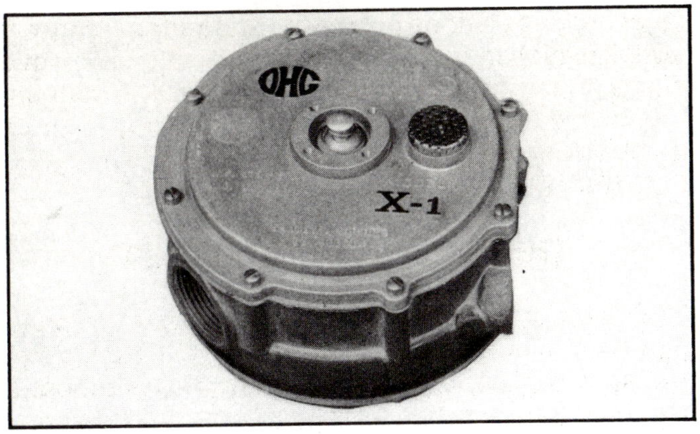

The OHG company, although relatively new, encompasses some of the most experienced engineering talent in the field of LPG carburetion. The first OHG product was this improved vaporizer, which has all the water fittings at one end and all gas fittings at the other (easing installation and plumbing).

stance that has been properly investigated, the propane system has demonstrated a very high resistance to damage. Campers have rolled off mountains, highway crashes have severely deformed propane tanks—without creating any leaks—and vehicle fires have even started from other causes without the propane system causing any additional complications. In the few cases where propane has been a contributing factor to damage or injury, *in every case, the cause has been traced to carelessness and/or negligence when the equipment was installed.*

This is the kind of safety record the propane companies and the carburetion companies would like to maintain—and the main reason they are looking to the fleet market for motor-fuel conversions, rather than the private passenger car. Professional fleet mechanics can be readily trained to perform safe propane installations and conduct proper preventative main-

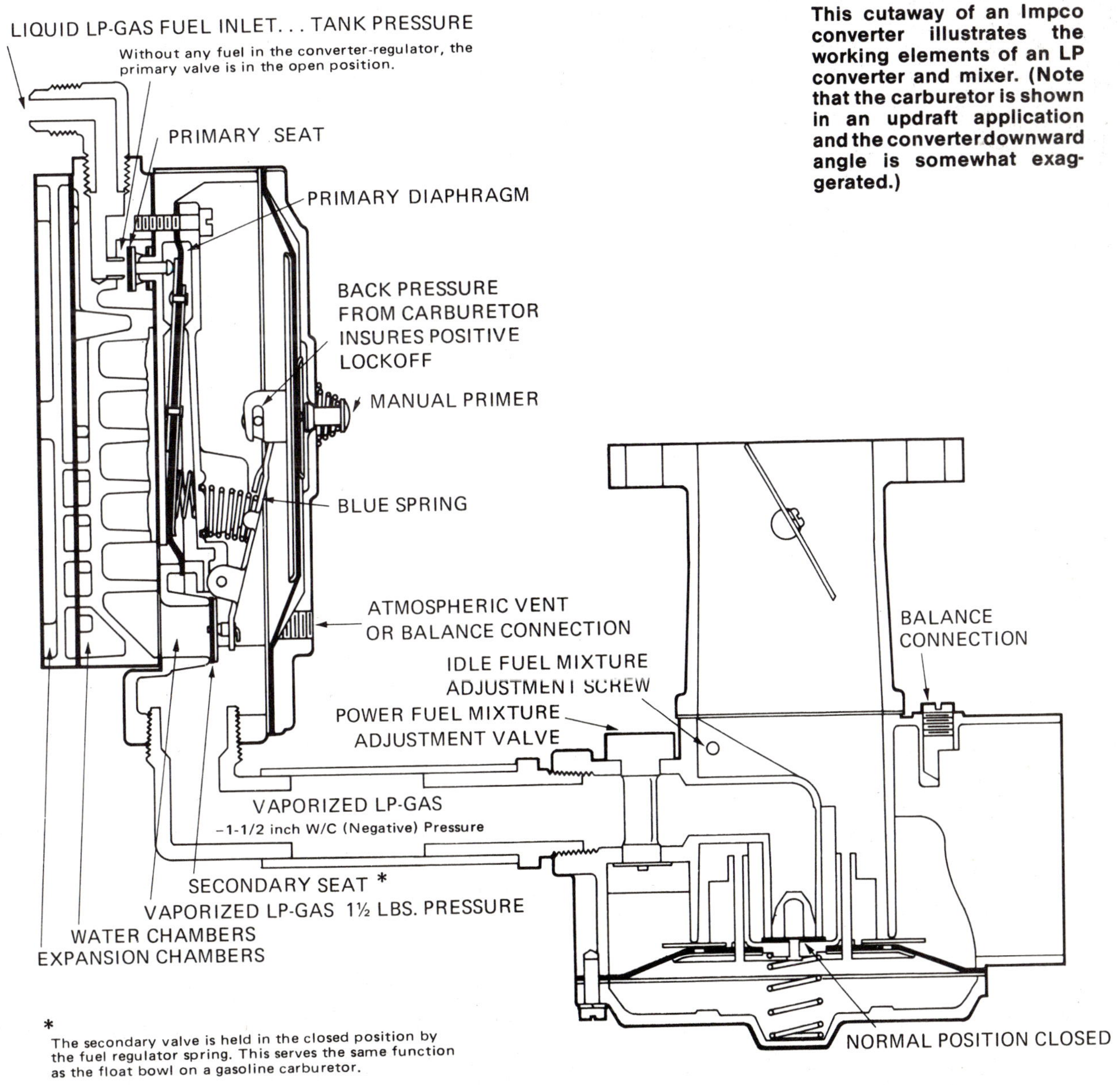

This cutaway of an Impco converter illustrates the working elements of an LP converter and mixer. (Note that the carburetor is shown in an updraft application and the converter downward angle is somewhat exaggerated.)

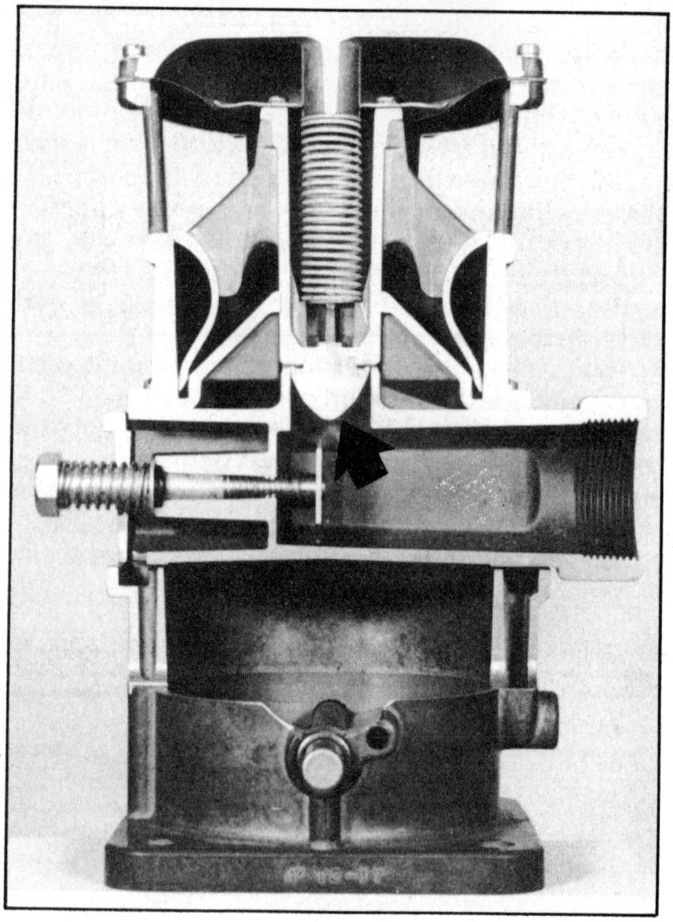

This cutaway of an Impco carburetor graphically illustrates the relatively few major components and simple design (as compared to a typical gasoline carburetor). The diaphragm (top) responds to engine vacuum or airflow and draws up the air valve; as the air valve is lifted higher from the seat, more fuel is supplied. The shape of the gas jet (arrow) controls the ratio of fuel to airflow. Screw at left is power adjustment.

tenance of the fuel equipment. The backyard mechanic and the untrained gas-pumper at the corner gas station are not as reliable, and most companies discourage independent privateers. It would only take a few sloppy backyard installations, which could conceivably lead to dangerous fire hazards, to create a very bad public impression for the industry and start an irreversible—and totally unjustified—furor over the "dangers" of propane fuels.

In 1977, several fires in buses belonging to the Santa Clara County (California) Transit District were played up in inflammatory newspaper headlines. One emotional bureaucrat described the buses as "mobile crematoriums for public mass transit fueled by explosive propane." After lengthy investigations, no less than six government agencies concluded that the propane systems had nothing to do with the fires and, in fact, had not ruptured or added to the fires, despite the fact that one of the buses had almost completely burned up (the tanks were intact and containing propane after the fire). As it turned out, the fires had been caused by inadequacies in the engine design and lack of proper maintenance on the engines, which resulted in overheating problems. Unfortunately,

The latest OHG project is a totally new carburetor design (something that comes along, it seems, only once every 25 years). Though a prototype is shown here, production models should be available in the near future. The OHG mixer accepts standard Holley base-plates (4-barrel) and flows about 100cfm more air than comparable brands. This new design provides considerable adjustability (for any application), and disassembly/assembly is simple and straightforward.

many of the residents of this county will remember the sensationalism created by the news media and the conclusions of the unbiased investigators—which totally vindicated the propane fuel systems in question—was not reported with equal fervor.

The propane industry tries very hard to avoid this kind of "bad press," which discourages most individuals from buying and installing propane equipment on privately-owned vehicles. An installer doesn't really have to be an engineer or a genius to do the job right; but it does require a certain degree of safety-consciousness and common sense—and history has proven that not many individuals possess

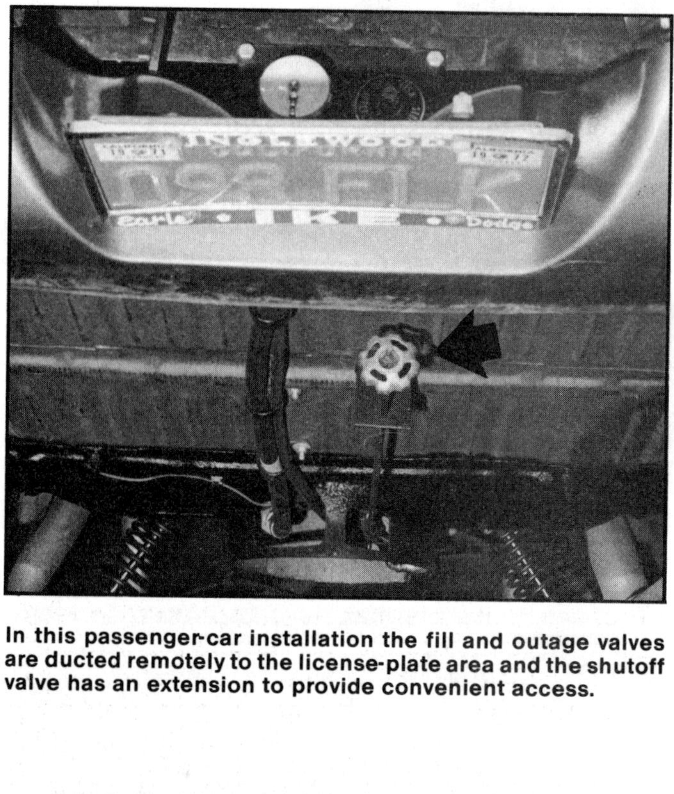

In this passenger-car installation the fill and outage valves are ducted remotely to the license-plate area and the shutoff valve has an extension to provide convenient access.

When a single large tank won't fit, "manifold" tanks are used: two smaller cylinders connected by welded steel tubes.

these qualities. Any of the propane companies can relate numerous horror stories about "backyard installations"; and though most individuals mean well, there is no substitute for experience, proper tools, and proven techniques.

Propane equipment safety was the subject of a thorough study in Holland by the Research Institute for Road Vehicles (TNO). The institute installed propane systems on a Renault 12, a Peugeot 504, a VW Passat, and a Fiat 128 and conducted a series of head-on and rear-end collisions with a 400-ton concrete crash barrier. In every crash, the propane systems came through without causing any fire hazards; and most crashes did not even cause any minor leaks in the systems. The Renault 12 hit the barrier head-on at 36mph and was hit from the rear at 30mph; the Peugeot 504 was hit head-on and from the rear-end at 30mph; the VW Passat hit head-on at 36mph and was rear-ended at 30mph—all without any notable damage to the propane systems!

In 1978 the institute performed an additional series of tests, and the results again showed the rugged durability and safety of propane components. In these tests, the vehicles included an Opel, a Ford Taunus, a Volvo, and some Leyland/Verheul passenger buses. The Volvo was involved in an incredible test in which it was struck consecutively, front and rear, by two speeding railroad *trains*. The propane tank was the only component of the car to survive the crashes intact. In another test, a propane tank was bolted to a concrete barrier and a wedge-shaped steel prow was rammed directly into the tank at 12mph. The tank survived without any evidence of leakage. The same test at 19mph almost flattened the propane tank; and although there was some subsequent leakage from the filling valve, the tank survived intact. When a Ford Taunus was driven into a barrier-mounted tank at 24mph, there were no leaks and only minor deformation of the tank.

Propane tanks are manufactured from 1/4-inch steel to rigid specifications dictated by the National Fire Protection Act (NFPA). This construction makes them very strong—much stronger than a sheetmetal

If you don't need the locking feature of a standard remote-fill box, the fill and outage valve can be ducted to the stock (gasoline) fill location—either on the fender or behind the license plate.

gasoline tank; and the containment pressure in the tank adds to the strength, making them even more resistant to deformation or puncture. In addition, since the tanks are generally round in shape, sharp objects are deflected away from the tank surface in a crash.

A propane tank also has several built-in safety features. All tanks must be fitted with an excess-flow valve and a pressure-relief valve. The tanks are built to the ASME (American Society of Mechanical Engineers) boiler and pressure-vessel code and can withstand pressures in excess of 350psi (the typical working pressure is only 150 to 190psi), with the pressure-relief valve set for 250 to 312psi pressure. If anything happened to cause the pressure in the tank to raise above this level, the relief valve would open and allow enough propane to escape to bring the pressure below the pre-set level.

In the case of a fire, this vented gas could be ignited, but it would only create a spurting flame for a short period of time. *It would not cause an explosion.* And for this reason, the pressure-relief valve is designed to be ducted to some point outside the body; and in the event that it must vent pressure, the propane is directed away from the car. Once it is a few feet away from the source, the vapor has dissipated enough that it won't combust (propane's limits of flammability in air are 2.37 to 9.5%, so even rich mixtures of propane won't ignite).

The second important tank-mounted safety device is the excess-flow valve, which is inside the valve for the engine fuel line. If for any reason a line ruptures in the engine compartment—due to an accident or any other minor mishap—the excess-flow valve will immediately sense the excessive rush of fuel and shut off the liquid supply at the tank. The only fuel that is free to escape is the little bit left inside the fuel line. If there is any sort of fire after the rupture, there is really very little free propane that can possibly add to the hazard.

Downstream of the tank and somewhere in the fuel line is another device known as the hydrostatic-relief valve. This component is installed in the fuel line and ducted outside the body of the car, usually underneath and away from exhaust-system components. The hoses used for liquid lines in a propane conversion must be UL-approved LPG hose, built to hold a working pressure of 350psi and a burst pressure of 1750psi. However, before a hose could reach this pressure and burst, the hydrostatic-relief valve would open, relieve the pressure, and cool off the hose. It's designed for a situation where the engine is shut off and the tank valve is closed, leaving the tank-to-engine hose full of liquid. If the pressure in this line should rise—due to heat or a fire—the relief would dump the pressure and prevent the line from bursting.

When these devices are combined with the already-mentioned filter/fuel-lock and the redundant safety features of the pressure regulator and the carburetor, a propane fuel system becomes safe enough for even Ralph Nader; however, the entire system must be properly engineered and installed in a totally professional manner. And it must be stressed one more time that this last qualification is crucial!

PROPANE REFUELING

When you pull into the propane station, you may say, "Fill 'er up." But he won't...not quite! Propane motor-fuel tanks are not intended to be filled completely; some room must be left at the top of the tank to allow for vapor expansion. If the tank were totally full, the pressure-relief valve could possibly blow open whenever the car was parked in the sun (the absorbed heat might raise the internal vapor pressure above the preset relief level).

Propane tanks are stamped by the manufacturer to indicate the total internal volume (gallonage); but this number is followed by a "W.C." symbol, which stands for water content. A tank that is stamped *42 W.C.* means that it will hold 42 gallons of water; but *in actual practice, this tank will hold only about 80% of this total gallonage*—about 34 gallons of liquid propane.

Manufacturers make certain that the tank will only be filled to 80% of capacity by incorporating an "outage valve" in each tank. The fuel-level gauges on propane tanks read in percentage of "fullness" (rather than how many gallons remain in the tank), and when the tank is being filled, the outage valve must be cracked open a little bit when the tank nears 80% full.

At the right of this bulkhead fitting is the *hydrostatic-relief valve*, which must be mounted between the tank and the engine. It will open momentarily (to vent excess pressure) if pressure in the supply line should ever exceed the preset limit.

A remote-fill hookup isn't necessary in most pickup installations, because the tank fill is readily accessible; but in the case of this camper, the fill and outage valves had to be extended outside the body. Popoff valves should also be externally vented.

The remote-fill hookup usually terminates in a lockable steel box (for security).

The valve will hiss a small amount of propane vapor until the tank nears the pre-set filling limit, at which time the outage valve—sometimes called the "20% valve"—will emit a small amount of white propane liquid. As soon as this occurs, the attendant will stop filling the tank and close the outage valve. (This less-than-full maximum capacity is another good reason to use the largest tank capacity possible.)

Propane refueling is really a simple, safe, and environmentally clean operation. The dealer will have a large propane storage tank full (almost) of liquid, under which is mounted a substantial electric pump built to take high pressures and cold temperatures. The heavy refilling hose leads from this tank and is terminated in a large brass fitting that the dealer screws right onto the filler valve of the vehicle tank. Near the screw-on connector are a gate valve, or "faucet," and a quick-action shutoff lever.

When the dealer has hand-tightened the fitting onto the tank filler valve (or the remote outside filler if the tank is mounted in the trunk of a passenger vehicle), he will switch on the electric pump, set the gallonage meter to zero, open the gate valve and flip the shutoff lever. Liquid propane is now pumped into the vehicle tank until the outage valve, which the dealer will crack open when the tank gauge nears 80%, emits liquid propane. At this time, the dealer will shut off the tank outage valve, close the shutoff lever and turn off the electric pump. When he returns to the vehicle, he will then close the gate valve on the hose and unscrew the fitting from the tank. During this operation, the dealer will probably be wearing gloves because the small amount of propane that is in the refill hose (only that which is between the gate valve and shutoff lever) will be released and the ultra-cold propane could burn his hands.

Other than the negligible amount of propane that is emitted from the outage valve, the small residue from the refill hose is the only hydrocarbon that escapes during the refueling process. It's estimated that during the manufacturing and transfer of propane, the amount of hydrocarbons released is 40% less than what is released in comparable open-air operations with gasoline. Have you ever watched a gasoline tanker truck fill the tanks in a service station? So much hydrocarbon vapors are escaping into the atmosphere that you can actually see them, like heat waves from hot pavement. This is one of the primary reasons—along with lower exhaust emissions—that propane-powered forklifts are used in enclosed factories and warehouses. (Propane-fueled lifts are practically mandatory in food-handling warehouses and ships' holds because they can be refueled by simply switching an empty tank for a full one, eliminating virtually all unwanted hydrocarbon emissions.)

Because of the gaseous nature of propane, there are some peculiarities that users must be aware of. One of the most obvious is the difficulties encountered when trying to refill a tank that has been

The valves required on an approved LPG motor-fuel tank are (from left to right): *fuel filler*, where fuel is put into the tank; the *outage valve*, with round warning tag; a *vapor fitting*, used only when the vehicle has an additional need for propane vapors (such as a camper that uses this line to supply the stove or refrigerator); the *pressure relief valve*; and the *liquid valve*, which supplies fuel to the engine.

All legal LPG tanks must comply with strict structural regulations. The water content is also shown on the tank specifications. This indicates the total volume of the tank; but remember, when filling with LPG, the tank will only accept 80% of this capacity.

Inside the tank, the *liquid valve* is ducted to the bottom of the tank; and the *vapor valve* is always above the liquid level. The float activates a fuel-level gauge on the outside of the tank.

As with all flammable fuel installations, safety is always of the utmost importance. Absolutely no other hose should be used than that specifically marked to take the pressure of liquified petroleum gas.

heated. If the tank is exposed to direct sunlight or mounted in the trunk of a dark-colored car that has been in the hot summer sun all day, the dealer may have trouble filling the tank. The heat from the sun can raise the gas pressure inside the tank; and if the refilling pump is a little weak, it may not be able to overcome the internal pressure (because of a valve inside the filler that prevents over-filling).

In a case like this (which is not very common, except on really hot days), the only choice is to reduce the pressure in the tank. The best bet is to cool the tank with cold water and/or wet rags. After the tank has cooled some, it can usually be filled if the outage valve is open all the way to relieve the pressure in the tank. (Sometimes it is also helpful to crack the "vapor fitting" on the tank.) Once the pump overcomes the internal pressure and filling begins, the outage and vapor valves can be closed.

Most modern propane distributorships have excellent powerful pumps where this procedure won't be necessary; but if you encounter a weak pump when the weather is very hot, as happens occasionally, this technique usually overcomes any problems. One way to lessen this problem is to paint the propane tank a light color. Many tank manufacturers paint their tanks white, which can keep the tank 15 to 20° cooler; but some stock tanks are painted dark gray, and some pickup owners even like to paint the propane tank to match the color of the vehicle. If there are no refilling problems, there is really nothing wrong with this; but if refilling becomes troublesome, it would be wise to paint the tank white or devise some method to shade it from direct sunlight.

Some readers who are really dollar-conscious may be thinking that the ideal setup would be to have a large propane storage tank in the backyard, from which the vehicle could be refilled at any time. In theory this would be great, especially if the user could buy the propane at wholesale prices. However, even though many multi-vehicle fleet operations maintain their own refilling stations, this is not always practical for private individuals. Many states have very stringent "road tax" laws that apply to propane used as a motor fuel; and though most wholesalers will deliver propane in large quantities, (which usually requires a fairly large holding tank), if you use this fuel in your car or truck, you will be required to pay the requisite tax on every gallon.

THE FUTURE OF PROPANE AS A MOTOR FUEL

Some readers may be hesitant to spend money for a propane conversion because it is sometimes difficult to find the required components, and many manufacturers actively discourage private individuals from doing their own conversions. In addition, there is always the question of propane availability. After all, what is the use of converting to propane if it's just as difficult to get and just as expensive as gasoline?

As for the first question, given the excellent existent safety record of propane motor fuel—a record that the industry views with considerable pride—it is understandable that the propane establishment is reluctant to foster freelance conversions of private motor vehicles that may be done in home garages by unskilled personnel and maintained by drivers unfamiliar with propane motor fuel. In the past, major carburetion companies have largely concentrated their primary sales efforts on fleet operations (with mechanics that have been trained and proven fully qualified to operate and maintain propane systems) and to dealers already in the propane distributing business.

Industry leaders indicate that large fleet operations will remain the main thrust of most future marketing plans, but it is also possible that some general-purpose installation centers will emerge. The industry is beginning to take a serious look at domestic use of propane as an automotive fuel; and one of the major propane distributors, Petrolane, is currently studying the possibility of establishing a chain of installation centers around the country that will do conversions and provide suitable training for users.

And as far as availability goes, it is important to realize that propane is one of the few out-of-the-ground hydrocarbon fuels that is in excessive supply...worldwide. In many parts of the world where oil and gas are found, the refineries have had no facility for storing gasses and have simply "burned them off" into the atmosphere. The huge flames from the burnoff stacks in the great fields of Saudi Arabia could be seen by orbiting astronauts. And thankfully, even the oil-rich Arab states now realize the finite quality of these resources; they are piping off "surplus" gasses and storing them in new cryogenic facilities built with Japanese technology.

As mentioned previously, Canada, Australia and New Zealand have tremendous gas reserves; and there are some large, untapped fields in this country as well. We have considerable reserves of gas in Alaska, (though it may be some years before we can utilize those supplies). Currently, new wells are com-

THE PROPANE FUEL SYSTEM

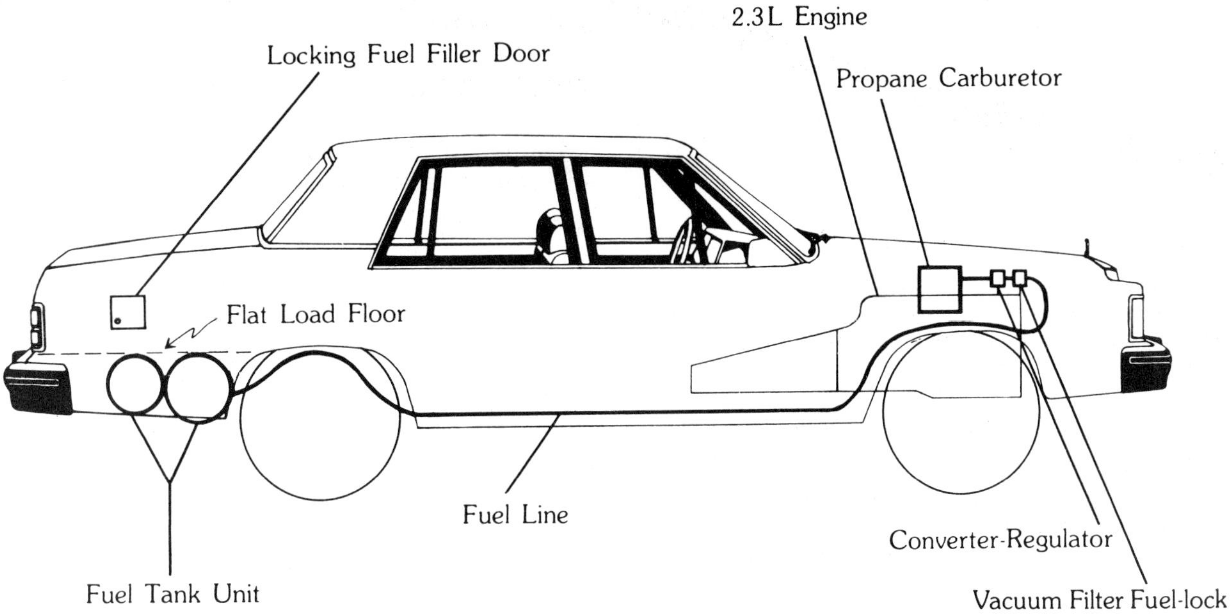

Factory engineers are fully aware of the benefits of propane fuel. A number of commercial trucks have been offered with LPG options and in recent years Ford has produced factory conversions for high-mileage fleet users. This propane-fueled Ford Granada is not simply a production vehicle fitted with propane carburetion. It has been completely re-engineered to take full advantage of the LPG. The 2.3-liter four-cylinder engine features high compression, special rings, altered ignition, blocked heat risers, low-temp thermostat and revised timing. With multiple tanks under the trunk, these cars have 35% greater range than their gasoline-powered counterparts.

ing in regularly in the Philippines. Pemex, Mexico's national oil company, has expanded proven and probable reserves in 15 Mexican sedimentary basins (plus recent discoveries in the Gulf of Campeche) to 200 billion barrels; and in the recent past, Pemex announced that it expected total production of liquid hydrocarbons of more than 2.25 million barrels a day (most of it supposedly slated for export to generate the capital required by Mexico's proposed industrial development program). Despite some fumbled politics, we are still the most likely customers for Mexico's future exports; and there has been some effort to develop direct pipelines between Mexico and the U.S.

So in the foreseeable future, the supply of propane should be abundant; and most experts predict it will continue to offer a price advantage over gasoline (as a motor fuel). If there is a problem in the future, it lies in the storage and distribution. One industry expert predicts that it will be from 5 to 10 years before there are adequate cryogenic, storage, and distribution facilities around the world. As one of the leading world marketplaces for propane and natural gas, the United States is far ahead of other countries in such facilities; but they don't do us any good if the sources we import from aren't equally prepared. While the U.S. Department of Energy (DOE) analysts are predicting that there are enough worldwide supplies of propane to meet U.S. needs, the political side of the DOE is still uncertain about U.S. energy quotas and whether we should continue to limit U.S. imports to 1977 levels. An overall demand increase of 36% in domestic propane usage is expected over the next 10 years; but with domestic production flat, we can expect an increase in our propane deficit of 600%! The deficits are made up for with imports.

A source at Dutch Shell has indicated recently that the Arab nations are now considering how to capitalize on their newly-acquired stores of hydrocarbon gasses. One possibility is that they might *require* their existing customers to take supplies of gas with every oil purchase. This would have a dramatic worldwide effect on the supply and distribution of propane and natural gas because the countries who are totally dependent on the Arabs for crude oil would find themselves with sudden supplies of gas that they would either need to export or find uses for in their own countries. Increasing their domestic usage of the gasses would most certainly mean increased use of propane as a motor fuel.

The 1980s is proving to be an interesting and exciting decade in worldwide usage and conservation of energy, and clean-burning propane will very likely become a bigger part of our lives—especially whenever and wherever energy usage conflicts with environmental considerations. And all things considered, it could be the fuel of *your* future.

CHAPTER 2

BUILDING AN ENGINE FOR PROPANE FUEL

An engine in good condition will run excellently on propane or dual-fuel; but if you're starting a rebuild, you can take further advantage of modifications that optimize an engine for LP fuel.

BUILDING AN ENGINE FOR PROPANE FUEL

The plain fact is: converting to propane fuel isn't going to make a "smoker" into a blue-ribbon motor. The greatly extended engine life commonly associated with propane fuel just does not apply when the engine has already given up most of its useful life (on gasoline). To obtain the fullest advantage, the conversion should be made on a fresh engine; but in addition, it is ideal if the engine has certain specific modifications to fully utilize the benefits of the fuel.

This isn't to imply that a gasoline engine with low mileage isn't suitable for a conversion. Virtually any sound engine (tight rings, good valves and seats, good valve guides, etc.) will function efficiently on propane. However, if the engine has less running time on gasoline, it should have a relatively longer operational life after the conversion.

We'll make a lot of specific recommendations in this chapter to help readers *optimize* an engine for propane. Most of these tips are easily accomplished without a complete teardown; but if you *are* rebuilding an engine, you can start with a clean slate to take maximum advantage of gaseous fuel.

THE IGNITION SYSTEM

No single aspect of a propane-fueled engine is as important as the ignition system. According to one industry expert, more than 85% of the problems of fleet and individual conversions can be traced to the ignition; and more than a few engines have been ruined by careless attention to the ignition requirements!

Why should ignition requirements be any different for a propane engine than for gasoline? Let's look first at the job done by the spark plug, the final link in the ignition chain. Before a spark plug can fire a fuel/air mixture in the combustion chamber, the air space between the electrodes must be *ionized.* That is, the electrons in the air space are "aligned" in such a way as to build an electrical "bridge" that permits voltage to arc across the gap. This ionization depends on the relative quality of the mixture in the chamber (and a number of other factors); and to a degree, this determines the *voltage requirement* for ignition to start in the cylinder. The ignition coil develops the voltage (the secondary voltage) required to bridge the spark plug gap; but when the voltage requirement is higher than the *available voltage* from the coil, the cylinder misfires.

Propane fuel is, in effect, less volatile than gasoline, making it tougher to ionize the gap. This raises the voltage requirement to initiate ignition in the chamber. Therefore, an ignition system that might otherwise work fine with gasoline may not work on a propane engine. In fact, experience has shown that even the best gasoline-oriented ignitions usually require some work to make them suitable for gaseous fuel.

One method for reducing the voltage required to ionize a plug gap is to simply make the gap *smaller.* When a conventional gasoline-engine ignition is used with propane, the plug gaps should be held to 0.025- to 0.028-inch. However, the gaps can generally be enlarged slightly—to the range of about 0.028- to 0.035-inch—if the ignition is one of the so-called "high-energy" systems. Most of the late-model engines with factory high-energy ignitions (designed for lower emissions) operate with plug gaps that are relatively wide, and the recommended plugs are intended for wide gaps (the side electrode is relatively far from the center electrode). Therefore, it may be difficult to adjust these plugs properly (keeping the side electrode parallel to the flat tip of the center electrode) for the

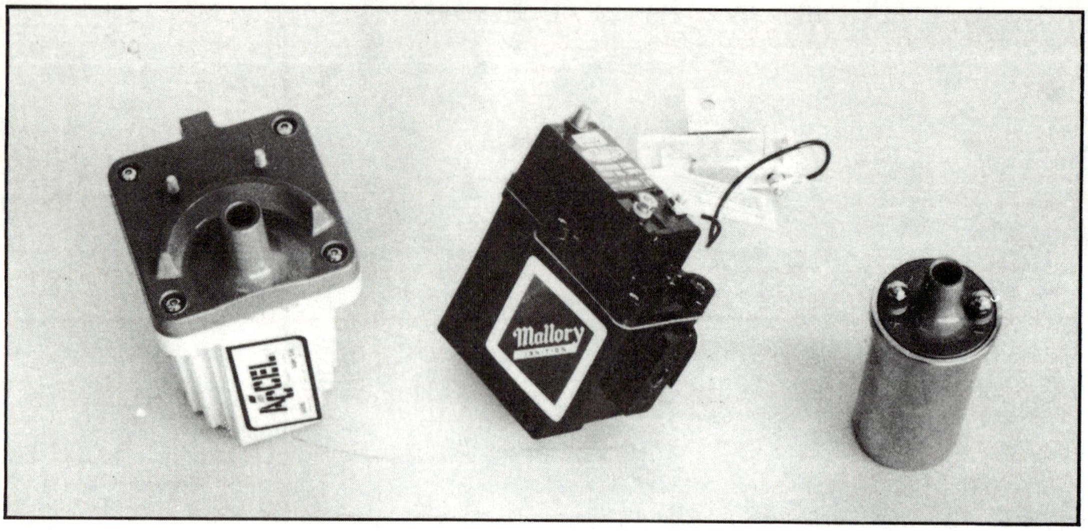

Regardless of whether you rebuild the engine completely for LP or convert a used engine, the ignition system must be in top shape. In most cases it will be better to replace the stock-type coil (right). The primary and secondary terminals are generally too close together, promoting voltage leaks. The Accel coil (left) has a "fence" to separate the terminals, while the Mallory has the terminals on separate sides of the case.

relatively small gap that propane requires. In such cases, check the spark-plug manufacturers' cross-reference sheets for a plug of the same heat range that can be readily gapped to the smaller specification.

Experienced gasoline mechanics know that richer fuel/air mixtures require less ionization voltage, but don't make the mistake of trying to enrichen the mixture of a propane engine to compensate for a weak ignition system. Though this will work with gasoline (under ideal conditions, such as low altitude and warm weather), propane engines don't respond to mixture changes in a similar manner. It is important to remember that the voltage requirements for propane are higher under *all* conditions.

Spark-plug *heat range* also affects the ionization voltage requirement; a hotter plug requires less voltage. Unfortunately, propane, which has less latent cooling effect than a liquid fuel, requires a colder plug. Actually, the heat-range requirement on propane will depend on the loads imposed on the engine. Lightly-loaded passenger-car installations can usually get by with plugs that are about the same as recommended for a similar gasoline-fueled engine; medium-duty usage generally takes one range colder than stock; and heavy-duty applications or turbocharged engines often require two ranges colder.

To reduce the possibility of *available* voltage falling below the *required* voltage, the resourceful mechanic is tempted to install a "hotter" coil, and thereby increase the secondary voltage. We've all seen the advertisements for racey "super coils" said to produce 50,000 volts. However, most of these souped-up coils must be considered with some skepticism.

Voltage can be viewed as electrical *pressure*; think of it like water pressure and compare the ignition system to ordinary household plumbing. If the voltage is increased and the rest of the ignition components are the same as used with the stock 20,000-volt coil, the additional pressure (voltage) is going to cause "leaks" to spring up all over the plumbing! When this happens, the extra voltage leaks away from the system (through small ground shorts) and is simply wasted. (If you look carefully on a damp, dark night at an idling engine that has been carelessly modified to produce very high secondary ignition voltage, this leakage looks like a scaled-down rendition of the Northern Lights.)

Though a better coil is certainly needed for a propane engine, it must be of the right design; and all of the other system components must be upgraded to handle the additional voltage. First of all, disregard any aftermarket coil that looks similar to a stock coil. The primary terminal is too close to the secondary terminal on this type of coil; and no matter how much wire is inside the coil (to increase the step-up ratio), the extreme voltage can easily leak from terminal to terminal without ever reaching the plugs.

Two of the best aftermarket coils for a propane engine are the Mallory Voltmaster Mark II and the heavy-duty Accel Super Coil. In the former unit the primary and secondary terminals are on different sides of the rectangular case; and in the Accel coil there is considerable separation between the terminals. Both coils have higher spark energy than stock coils.

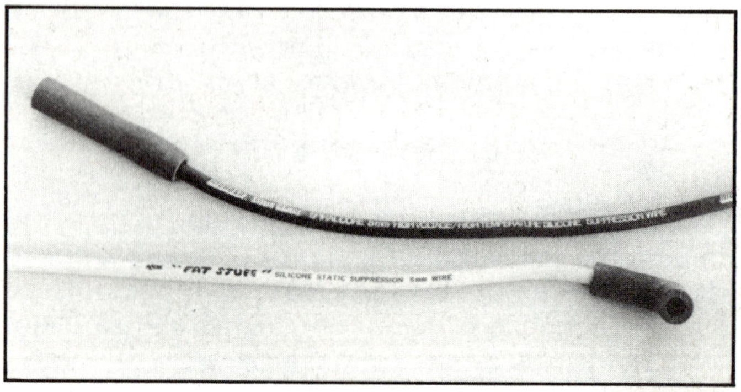

Once you boost the available voltage with a high-performance coil, the rest of the system needs upgrading to handle the juice. Silicone-jacketed wires, such as these specialty Accel and Moroso wires, are available through high-performance outlets and several automotive parts houses.

Thanks to a proliferation of aftermarket ignition products, there are plenty of heavy-duty components to handle the higher voltages. Most stock distributors—whether point-type or electronic—should serve well if in good condition; but the cap and rotor should be replaced with modern, high-dielectric (non-conducting) alkyd components. These are available in most auto-parts stores or accessory shops and are usually identifiable by their bright, distinctive colors; just ask for "heavy-duty" caps and rotors.

Fortunately, the late Ford, Chrysler, and General Motors high-energy systems work well with propane. Most of these recent engines already have high-quality alkyd caps and rotors. For example, the GM HEI system (for High Energy Ignition) is currently one of the best factory systems for both unleaded gasoline and propane. The unusual cap-mounted coil is as potent as many aftermarket coils; and, incidentally, the coil can be removed from the GM distributor and used on almost any engine. Some sort of mounting bracket must be fabricated to support the coil, but this is not a difficult job.

Though these stock systems are basically sound for maximum performance, most of them will benefit from some ordinary performance tuning tricks. One of the shortcomings of the GM system is that the rotors don't last as long as the rotors on conventional systems (due to the high voltage load). Only the best-quality replacement rotors and caps should be used with this type of system. It is also highly recommended that the two stock rotor hold-down screws (metal) be replaced with high-strength nylon screws (designed specifically for this purpose) available at most auto-parts stores. This eliminates the possibility of high voltage in the cap arcing to the screws.

Plug-wire routing is also important! Ignition systems with high secondary voltage have a strong tendency to "leak" voltage. The souped-up secondary voltage will very readily short to any potential ground source that offers less resistance than the relatively high resistance created by the spark-plug gap. Careful routing of high-quality plug wires through non-conducting wire supports is essential.

Every effort must be made to keep the wires separated from one another. The voltage in one wire can affect other wires that are nearby, even if they are not physically touching. This usually happens when the engine is highly loaded (just when it can do the most damage), and it is most likely to occur between two wires (cylinders) that are close together and sequential in the firing order. The phenomenon is called cross-fire; and the most common culprits are cylinders #5 and #7 of Chevrolet and Chrysler V-8 engines, cylinders #7 and #8 of Ford 302 and 460 engines, and cylinders #5 and #6 of Ford 351-C and 400-M engines.

It doesn't require much imagination to guess what can happen to a hard-working engine when the spark plugs fire out of sequence. Special attention must be paid to keeping the problematic wires physically separated as much as possible. It is also important to use the highest quality silicone-jacketed spark-plug wires. The latest Detroit ignition systems come with pretty good silicone plug wires and silicone plug boots. Such wires are a virtual necessity on a propane engine. Aftermarket companies also offer excellent replacement wires. The best wires have an 8mm insulation, and some suppliers are now offering wires with 9mm insulation. Reputable companies, such as Autotronic Controls (the MSD Heli-Coil wire has proven one of the best at preventing cross-fire), Accel, Essex (Mag-shield), Mallory, and Moroso, offer them.

Badly-designed insulating boots are one of the biggest causes of voltage leakage. Most of the aftermarket ignition companies offer silicone boots, and the boots used on some factory systems are also quite good. However, you must remember two things about these silicone boots and wires: 1) they stretch easily, and 2) they fit tightly on the plugs. Do not pull

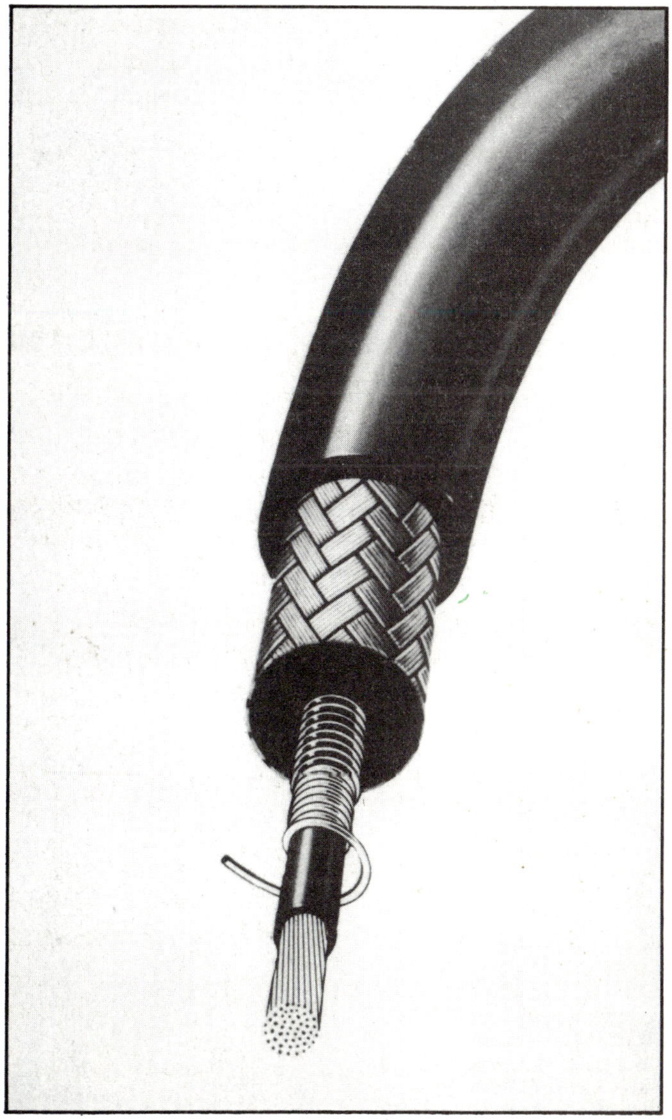

One of the ignition wires least susceptible to cross-fire problems is the Heli-Core wire by Autotronics Controls (developer of the MSD multi-spark ignition). It has a solid core (for strength) with a helically-wound radio-suppression conductor and a durable silicone outer jacket.

High-quality plug wires should also have rugged silicone plug boots. These boots from Mallory fit snugly around the plug insulator to prevent voltage leaks and they are large enough to be easily grasped and removed when the plugs need to be replaced.

the plug wires loose by grabbing the wire (this will stretch and possibly break the inner core that carries the current); pull them off by the boot.

In addition, when tight-fitting boots are installed, air is sometimes trapped between the plug shell and the boot; and when this air is heated by the plug, the boot, and possibly the wire may be blown off the plug. Some otherwise-intelligent mechanics have resorted to piercing the boots with an ice pick to release the trapped air, but this negates the whole purpose of the boot (quelling leakage). Mallory offers unique long silicone boots that don't fit quite so tightly on the plugs and don't trap air.

A final note on plug wires. Don't use so-called "high-performance" wires that feature solid-metal cores. With a high-energy ignition, these wires allow too much current flow and plug life is shortened (not to mention the distracting radio and TV static they produce). For most applications (short of an all-out racing engine), readily available silicone-jacketed, magnetic-suppression wires are best.

IGNITION ADVANCE

Even if you disregard the previous suggestions, one modification you cannot overlook is the distributor-advance curve. More head gaskets and pistons have been destroyed by excessive ignition advance than by any other cause, and propane has specific spark-advance requirements. Because of the high octane rating, it is commonly believed that

The "mechanical" ignition advance is controlled by the centrifugally activated mechanism inside the distributor. Altering the advance to suit LPG fuel can only be accomplished on a distributor test machine. Tuneup and performance shops usually have these test machines and an experienced professional can readily alter the distributor to provide the required LPG curve.

propane can withstand a lot of ignition advance without detonation. This is only partially true. Two factors seriously weaken the octane rating of propane: (1) the heat of the inducted air mixed with the propane and (2) the relatively high loads imposed on most propane engines. Therefore, under actual operating conditions—when the intake air temperatures are very high and the engine loads are high—the octane rating of propane may be reduced.

A slow burning rate gives a fuel a high octane rating, and propane has this at slower engine speeds; but as engine speed goes up and mixture turbulence increases, the burn rate speeds up. This lowers the useful octane rating! As a consequence, a propane engine can withstand a healthy load of *initial* ignition advance but must have less *total* advance than a gasoline engine.

How much total advance you can safely use depends on the loads imposed on the engine! For a heavy-duty application, such as a truck or a pickup that carries a camper or tows a trailer, *maximum total advance should not exceed 28-30°*. This restriction cannot be emphasized too heavily.

A good ignition-curve plot that works well with propane would have 10° initial advance on the crank (for good low-end performance). Distributor centrifugal advance should be about 9° (remember that one degree in the distributor equals two degrees on the crank), with full advance achieved by 2250 to 2500 rpm.

Unfortunately, most late-model smog distributors have heavy springs on the centrifugal-advance weights (to limit emissions of NOX), and generally lighter springs will have to be installed to ensure full advance by 2500 rpm. Ignition-advance modification

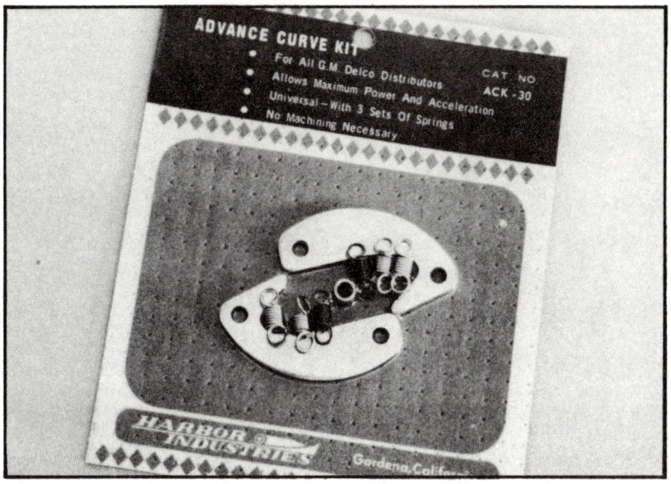

Advance curve "quickie kits" are available from many sources. These kits include modified advance weights and several sets of advance springs (with different tensions). When installed properly, these components alter the centrifugal advance, however a distributor test machine must be used to verify that the new curve does, in fact, match the recommended LPG specs.

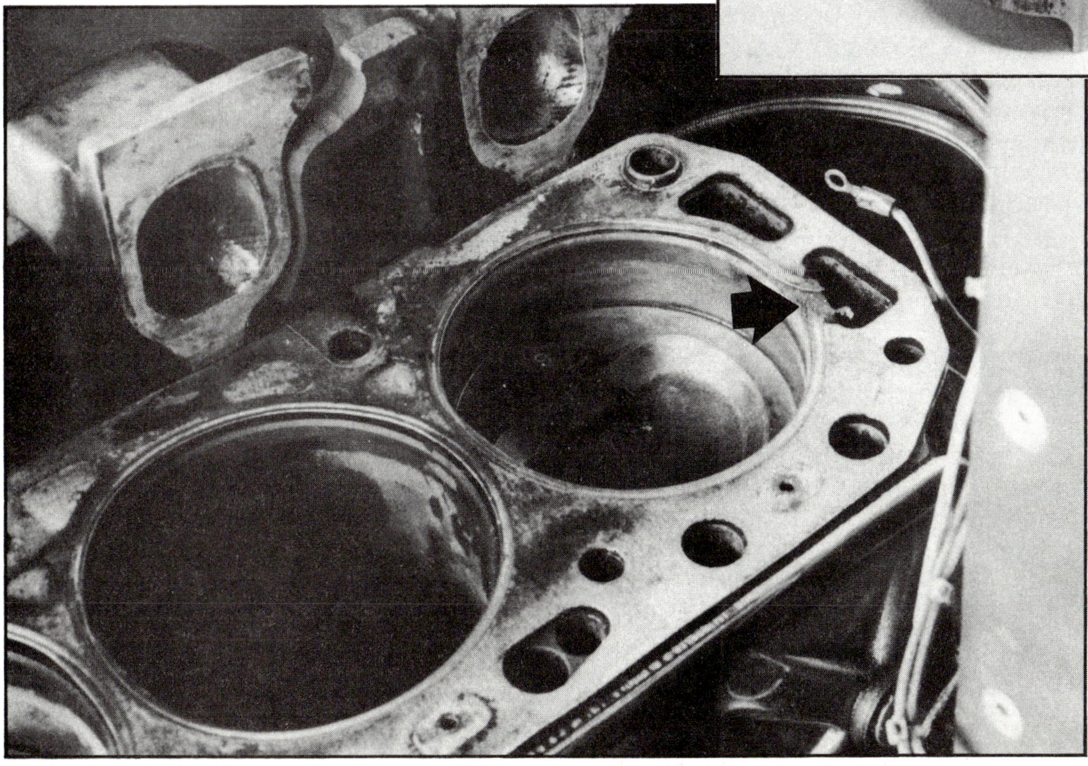

Excessive ignition advance can cause detonation, which dramatically increases combustion pressures and temperatures, until something gives—like the head gasket or a piston! Total advance on LPG fuel must always be limited to no more than 28 to 30 degrees.

Although the advance should increase more rapidly than stock, the total advance should be reduced; so the travel of the centrifugal mechanism must be limited. The length of the limiting slots (which determines the amount of advance pulled in by the centrifugal mechanism) can be reduced by filling them with braze and filing the slots to the required length; or a prefabricated limiting plates, like this Gann unit, can be installed.

kits (which include lighter springs) for most popular distributors are generally available in local automotive shops that sell high-performance equipment; but in addition to lighter springs, the distributor must be checked to determine how much total advance it delivers. In many instances, the distributor will allow *too much advance*; in this case, the advance mechanism must be suitably modified to reduce the total advance.

Some distributors have built-in advance "limit stops" that can be adjusted to gain a suitable propane curve. If this is not possible, one of the simplest alternatives is to install an advance limiter plate. Gann Products (of 9540 Stewart and Gray Rd., Downey, CA 90241) offers these plates in increments of 5°, 7°, 9° and 11°. The 9° plate would fit our sample curve exactly.

In any case, modification of the distributor-advance curve should be accomplished on an accurate distributor testing machine. Any local tuneup shop that has such a machine should be able to modify a distributor to suit the above specifications. Of course, in some states, the emissions laws prohibit this practice on late-model passenger vehicles (under 6000 lb GVW); *but some regulations may also exempt propane conversions.*

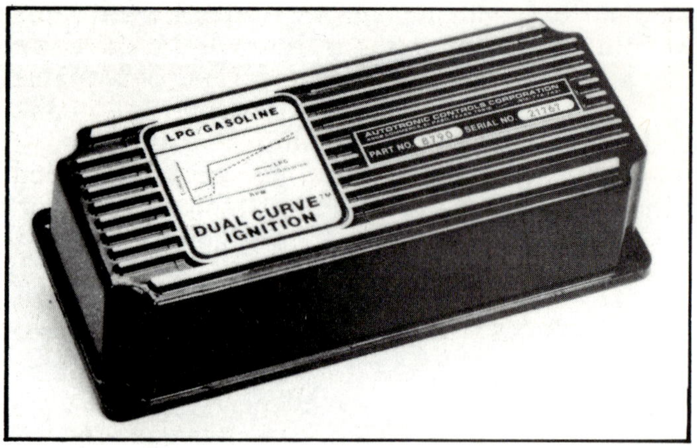

Autotronics Controls offers a special ignition developed strictly for propane-fueled engines. It provides a high-energy spark and electronically alters the advance curve to suit the unique requirements of LPG. The module can also be switched (with a simple remote switch) to a conventional curve for gasoline operation. This system is ideal for dual-fuel conversions and is highly recommended for straight LPG engines.

The easiest methods of setting up a propane ignition is to use a special ignition package available from Autotronic Controls Corp. (6908 Commerce St., El Paso, Texas 79915). Called the MSD "Dual-Curve" ignition, this transistorized ignition will provide a suitable ignition advance for almost any conventional engine that is converted to LPG. The engineers at Autotronic Controls have developed circuitry that electronically alters the existing (gasoline) ignition curve (which has been optimized by the engine manufacturer) to produce a curve for engine operation with LPG fuel. This ensures compatibility and optimum performance with almost any engine (except the GM odd-fire V-6). Some users have even reported better LPG fuel economy after installing the MSD Dual-Curve box.

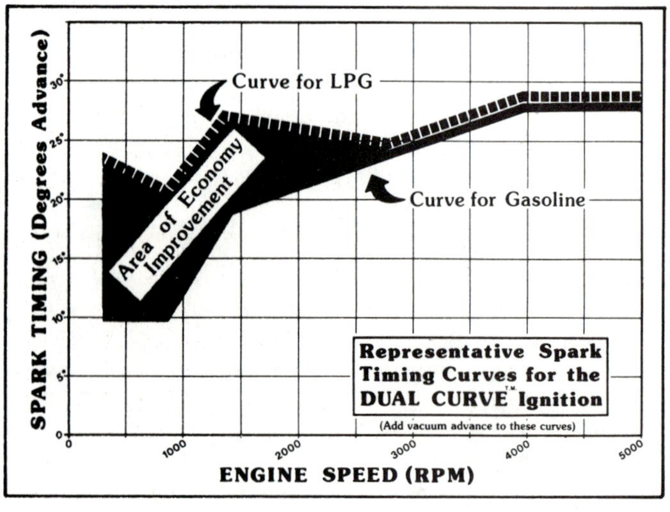

This is a comparison of gasoline and LPG ignition advance curves. An Autotronic Controls Dual-Curve ignition can be switched (electronically) from one to the other. Once the correct advance curve has been established, no further distributor modifications are required.

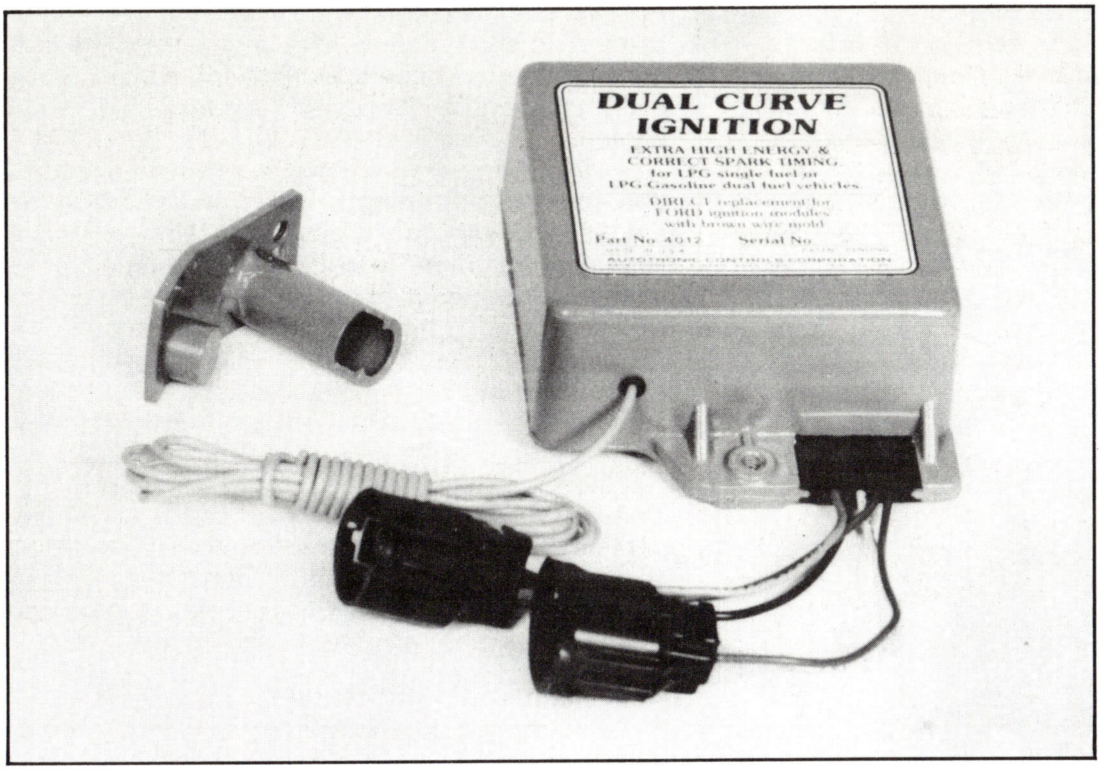

Autotronic Controls makes replacement modules that can be used on factory engines equipped with electronic ignitions. This is an MSD Dual-Curve for late Fords, which allows the ignition advance of these complex computer-controlled engines to be suitably altered for LPG fuel.

While the Dual-Curve can be used beneficially on any LPG conversion, it is especially useful on dual-fuel conversions. In the past, dual-fuel conversions (those that can run on either LPG or gasoline) have presented quite a problem. If the distributor is set-up for maximum performance with propane, it will not provide maximum performance when the engine is operated on gasoline. And likewise, if the distributor is set properly for gasoline, the performance on LPG will be compromised.

However, the MSD Dual-Fuel system can provide two separate (switch-selectable) advance curves—one suitable for propane and one suitable for gasoline. In addition, MSD offers a special Dual-Curve system for natural gas-fueled engines. This system is designed especially for fleet operators with CNG/gasoline dual-fuel vehicles. For ease and convenience, the MSD system can be easily wired directly to the fuel selector, so the correct ignition curve is automatically pre-selected, according to the position of the fuel switch.

The Dual-Curve ignition has another feature that makes it especially attractive to propane users. It has a circuit that automatically reduces the initial timing when the starter motor is activated. Typically, when the ignition provides a lot of initial timing (as is common in propane engines), the starter has a difficult time cranking the engine (especially when the engine is hot). The MSD ignition solves the problem by allowing the initial advance to be optimized for easy startup; and once the engine is running (and the starter switch/motor is deactivated), the initial advance is increased (electronically) to provide better low-speed performance.

It is also worth noting that conversion shops seem to favor the MSD Dual-Curve ignition because it is usually possible to simply mount and wire the MSD box (following the detailed instructions), and the system is ready to run on propane. On the other hand, if the stock distributor is going to be modified for propane operation, it must be removed from the engine, properly modified (a straightforward, but time-consuming task), and reinstalled. The MSD system eliminates this additional work; and it provides higher secondary spark energy, without requiring a special coil.

The *vacuum-advance* mechanism is also important on a propane engine. The vacuum advance is designed to provide additional spark timing for light-load cruising situations (when the mixture burns more slowly due to the reduced pressures, you need to start it burning sooner with more advance). A good vacuum-advance curve for most propane engines should begin increasing the advance at about 7 inches of manifold vacuum, and it should provide full advance by 10 inches vacuum.

On most engines, the total advance (vacuum and centrifugal) should not exceed about 40-46°; and with the centrifugal curve and initial advance previously mentioned, this would mean a total vacuum advance of 6-9° (which would be 12-18° on the crank). If the stock vacuum-advance mechanism does not meet this requirement (it must be checked on a distributor machine) it can be modified by (1) switching to another canister that meets the specs (the myriad of vacuum-advance cans available in this smog-engine era is staggering), (2) using an adjustable vacuum advance can (KEM Power System parts), or (3) by modifying the existing unit.

In the simplest case, the arm of the vacuum-advance unit can be drilled and pinned to limit the travel; but how do you know if the engine has "too much" vacuum advance? Follow our advance suggestions and you'll be in the ballpark; but if you notice that the engine "pings" at part-throttle and does not ping under load, the vacuum advance should be reduced.

Late-model engines have countless emissions solenoids, check valves, temperature valves, and vacuum-delay valves between the manifold vacuum source and the vacuum-advance canister. Only a factory service manual for the specific model in question can unravel these details, and such a manual should be checked before any modifications are undertaken. The shop manual should also have the part numbers and advance specifications of the various distributors and advance canisters available for the engine.

This MSD conversion for late GM vehicles with computer controls and oxygen-feedback sensors generates the correct electronic signals to allow normal operation with LPG fuel.

MSD also makes a Dual-Curve module for fleets that use compressed natural gas (CNG), which has slightly different requirements than LPG fuel.

As a final note on ignitions, remember that ionization voltages are going to be higher at higher altitudes, where the air is less dense. Not only is the voltage requirement higher, but the insulation of the secondary system is critical because the *air*, which was such a good insulator at sea level, is now at a lower pressure and voltage leaks develop more easily. It is possible to run a little more total advance at higher altitudes (about 4° more for every 5000 feet of altitude) because the mixture burns more lazily.

When properly tuned, a propane engine should have more power at higher engine speeds than a comparable gasoline engine. But if the weather gets a little chilly, remember that too much initial advance can make starting tougher. This may call for a little less initial advance and a comparable increase in the centrifugal advance.

REDUCING ENGINE HEAT

As mentioned before, a completely stock engine designed for gasoline may not be totally suited to propane. Most gasoline engines are designed to transfer some heat into the intake manifold for better vaporization of the gasoline (to "homogenize" it with air). However, this heat is unnecessary on a propane engine. The fuel is already a vapor by the time it leaves the converter; and the extra heat can substantially lower the octane value of the propane, leading to possible detonation.

One of the first priorities, therefore, in a propane

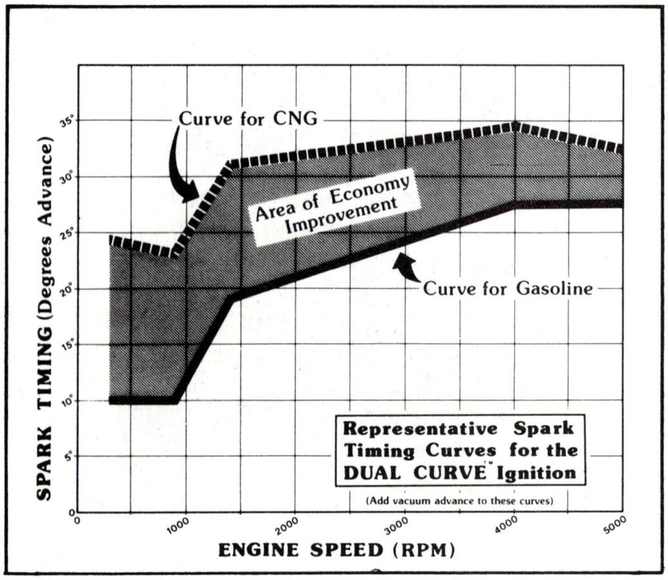

The relatively high octane of CNG allows more ignition advance throughout the curve than is typical with LPG fuel; as is shown by comparing this CNG Dual-Curve advance graph (recommended by Autotronic Controls) with the propane graph.

A remote air cleaner can increase power and forestall detonation by inducting cooler air to the engine. One horsepower is gained for every 10 degrees of air temperature reduction at the carburetor intake. A drop from 180° (typical underhood air temp) to 80°F is worth 10 horsepower.

High operating temperatures can radically alter the octane rating of LP fuel; therefore it is desirable to reduce underhood temperatures as much as possible. These heat-indicating tabs from Hayden Trans-Coolers can be helpful when tracking down heat sources. Marked in 150, 200, and 250°F increments, the spots will discolor to indicate the relative temperature. Much cheaper than gauges, these inexpensive tabs can be easily attached to the intake manifold, oil pan, transmission, etc.

conversion is to reduce the heat transferred to the intake. One of the easiest ways to accomplish this is to lower the operating water temperature of the engine by substituting a lower-value thermostat. In the last ten years, Detroit engineers have been forced to increase the operating temperature of most engines (to help meet emissions and mileage standards on gasoline); and production engines are now commonly fitted with thermostats that keep the operating temperature in excess of 200°F. However, replacement thermostats rated at 160°F are available in almost any auto-parts store; and this is a good starting point for a propane engine. It is also possible to find thermostats rated at 140°F, but they are not available for all types of engines.

And don't think, as some have in the past, that removing the thermostat altogether will reduce the engine temperature. Temperatures may go even *higher* when the thermostat is removed. Automotive cooling systems are designed to have a specific *rate* of coolant flow through the system with a conventional thermostat installed. Without one, the coolant flow may cause the water pump to cavitate, and this will substantially reduce the amount of heat that the system can carry away from the engine. So, always use a thermostat in any engine; but in a propane engine, use a thermostat that will effectively reduce the operating temperature as much as practicable.

A stock cooling system should be adequate for propane in other respects, provided that it is in good

Bill Chisholm uses a Century dual-fuel Accu-Carb on his 413cid Dodge motorhome. He eliminated detonation on long uphill stretches and substantially increased power by adding this huge air-cleaner housing (from GS & Company of Santa Clara, California) and ducting the intake with flexible hose to the front of the vehicle!

shape with no rust or scale present. It is not a bad idea to occasionally use a flushing solution to clean it, and always run anti-freeze or a rust inhibitor in the coolant.

Ultimately, the most important place to reduce the temperature is inside the intake manifold, and there are several other modifications that will help to accomplish this end. Another simple but effective measure is to get some flexible clothes-dryer hose and duct the snout of the air cleaner to a source of *cool* air.

On any engine, gasoline or gaseous fuel, the productive power increases about 1% for every 10°F that the intake temperature is reduced. This is especially important on propane because the advantages of the high octane rating can be offset if the intake temperature is relatively high—as is fairly common on conventional engines. Late smog-type engines may even have a flapper valve in the air cleaner snout that ducts hot air off the exhaust system to the air cleaner. This helps the engine during a cold start-up, but these systems can malfunction—causing the engine to induct hot air at all times. If possible, this valve should be eliminated on a propane conversion.

In general, there are three other sources that transfer heat to the intake tract: exhaust heat, oil heat, and water heat. The front and rear crossover passages of the intake manifold (on many V-8 engines) are used for water transfer between the cylinder heads, and there isn't much you can do about this (besides changing the thermostat). At least this heat is relatively remote from the carburetor; but right under the carburetor, there is usually a passage that ducts hot exhaust gases from one cylinder head to the other—the so-called heat risers. These should be blocked on a propane conversion. Most automotive shops that sell specialty equipment (hot-rod parts) carry special intake-manifold gasket sets that are designed to block the heat risers. It is also possible to use a small piece of thin stainless-steel sheetmetal to block the passages; these can be cemented over the riser openings when the stock intake gaskets are installed.

Most of us don't think about it, but on many engine designs where the intake manifold is also the valley cover over the lifter gallery, hot oil is constantly splashing onto the bottom of the manifold. So while

Ak Miller Enterprises offers this cold-air package with offset air-cleaner housing and 4-inch hose.

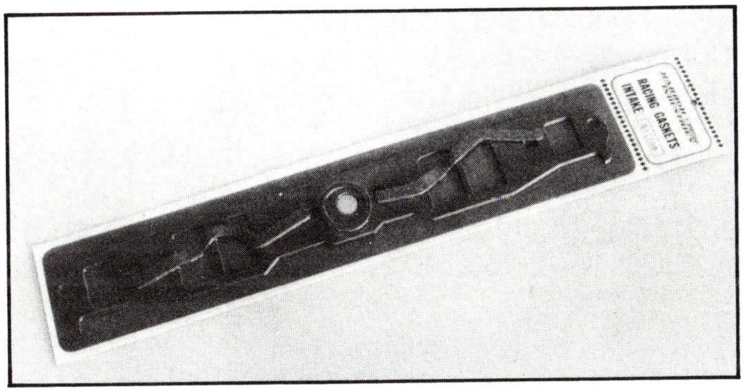

A straight LPG system can be optimized by blocking the heat-riser passages of the intake manifold with a set of racing gaskets. This is not recommended on gasoline engines because the engine will have cold-operation problems (the heat is required for adequate vaporization of the liquid fuel).

you're in the specialty auto shop, ask them about the type of oil-splash shields that are commonly used on racing engines. These are simple sheetmetal pans that fit beneath the intake manifold. They have metal clips that hold them in place, and they keep the hot oil from splashing on the underside of the intake manifold.

VALVES AND CYLINDER HEADS

Probably no other aspect of propane fuel is as misunderstood as *valve life*. If the average mechanic is asked about propane fuel and he at least knows what it is, he might answer, "It's a great fuel, but it destroys valves, burns pistons, and makes the engine run hotter." This guy is the one to shy away from when looking for someone to do a conversion because any problems that may arise from shoddy installation work will be blamed on the propane fuel. This is the same guy who, when confronted with a problem or failure on a vehicle that he has recently serviced, blames it on the manufacturer, saying something like, "All Chevrolets lose oil pressure like that," or "All of those Fords are hard on exhaust valves!"

What happens inside the engine when propane is combusted is no real mystery; the main difference between liquid fuels and gaseous fuels is that the latter do not create any latent cooling effects (through vaporization). They're already a vapor before they get to the combustion chamber. Gasoline provides a cooling effect on the exhaust valve during the overlap period that gaseous fuels do not; but by making the proper modifications, operating temperatures on propane can be kept below those commonly associated with gasoline—even in heavy-duty applications.

The valves are often the most vulnerable components in a spark-ignition engine. Excessive heat destroys the valve face, valve seat, guides, springs, and seals. Most modern engines are already running on the ragged edge in stock form; and confronted with an engine that exhibits signs of excessive exhaust temperature, the old-time gasoline mechanic is

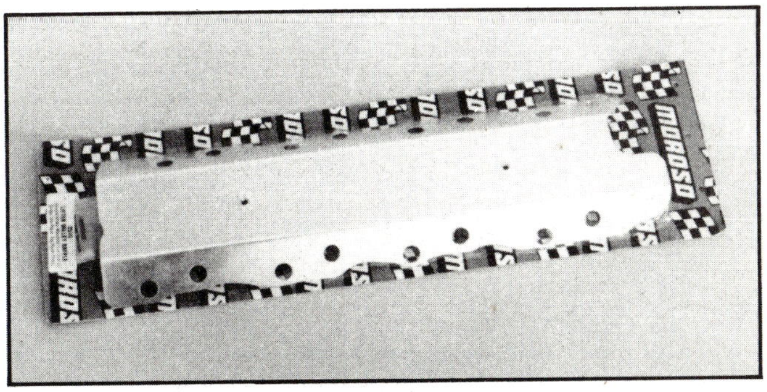

Another useful gadget available in most speed shops is an aluminum lifter-valley baffle. This simple device reduces intake temperatures by keeping hot oil from splashing on the underside of the intake manifold. If they don't have one for your brand of engine, it is simple to fabricate one from thin sheetmetal.

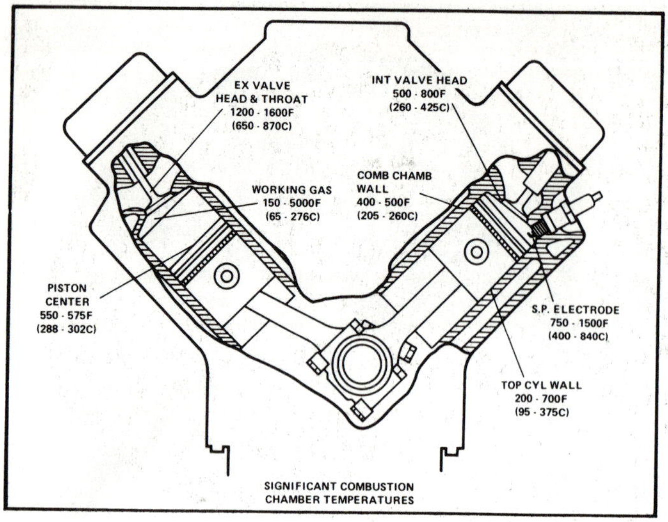

Operating temperatures inside an engine tell an important story. Latent heat of evaporation (of liquid fuel) significantly cools the intake temperature of a liquid-fueled engine. Gaseous fuels do not provide this benefit, therefore it is essential to reduce the intake temperature of an LPG engine by any practical means.

tempted to enlarge the carburetor jets and enrichen the mixture. But making a richer mixture will *not* reduce the valve temperatures in a propane engine like it does in a gasoline engine. The higher chamber heat of the propane engine requires more precise work in the valve/head area; and, as with the ignition system, borderline work that "gets by" on gasoline is often unsuitable for propane.

Some detractors point out that propane is an "unleaded" fuel and maintain that the lack of lead will promote early valve-seat failure (often called "valve recession"). It's true that propane does not contain any lead; and in an unmodified engine, this could certainly shorten valve life. But luckily, there are a number of ways to counteract this limitation.

Propane is very little different in this respect than unleaded gasoline, and modern engines designed for unleaded gas are perfectly suited for propane fuels (encompassing virtually all Detroit engines since 1975). The bulk of the heat put into a valve goes out through the valve face to the seat in the head, concentrating considerable heat in the narrow contact area between the face and the seat. With leaded gasoline, the lead deposits coat the contact area and prevent the hot surfaces from "spot welding" together. Without the protective lead, the surfaces can weld together when the valve is on the seat; and then the welds are torn apart when the valve is raised from the seat. This constant welding and unwelding is like a continuous grinding process that readily wears away the faces and seats.

Engines designed for unleaded fuels are equipped with better exhaust-valve materials and special "induction-hardened" valve seats that reduce the welding/grinding process. These countermeasures are equally as effective in a propane engine.

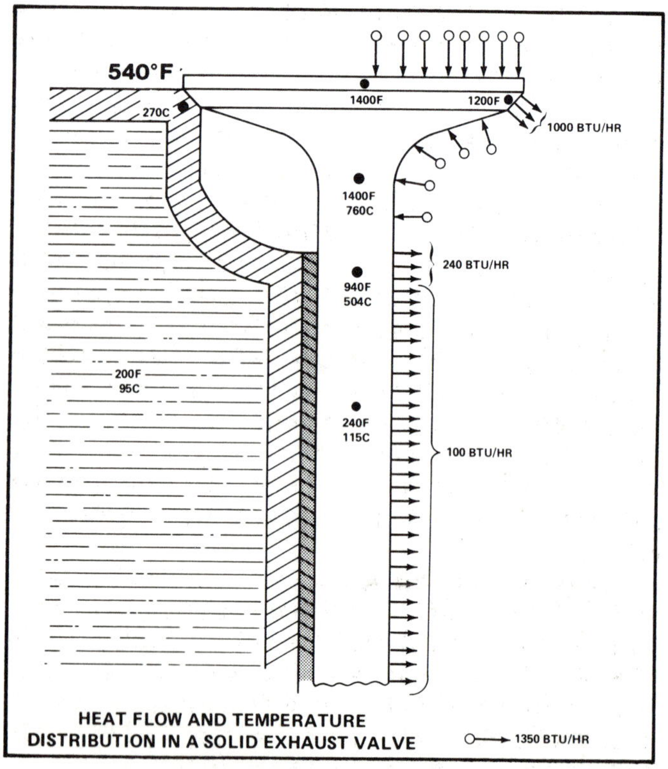

A typical engine will suffer some damage if the indicated exhaust-gas temperature exceeds 1400°F for more than a very short period. This diagram indicates that most of the severe heat concentrated in the valve is transferred to the cooling system through the small and intermittent contact of the valve head with the seat. Consequently, accurate valve work is critical.

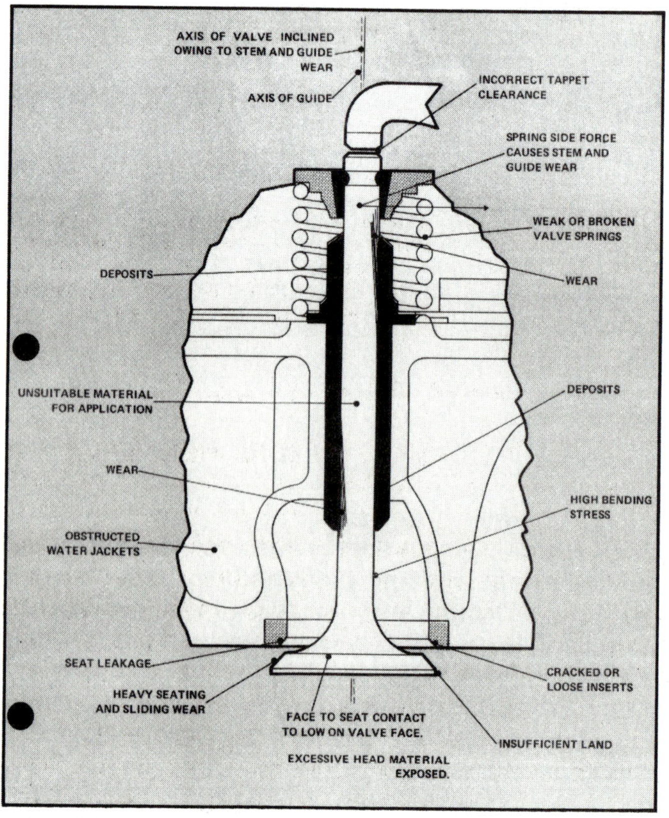

The valves are among the most highly stressed components in an engine. This list indicates common valve problems, and it is important that these potential difficulties be checked and corrected when rebuilding heads for a propane engine.

A precision valve job requires experience and conscientious dedication to details, and the propane engine demands the highest standards in head work. Conventional rebuild shops are often not equipped for this type work; look for a reputable shop that specializes in racing or performance preparation.

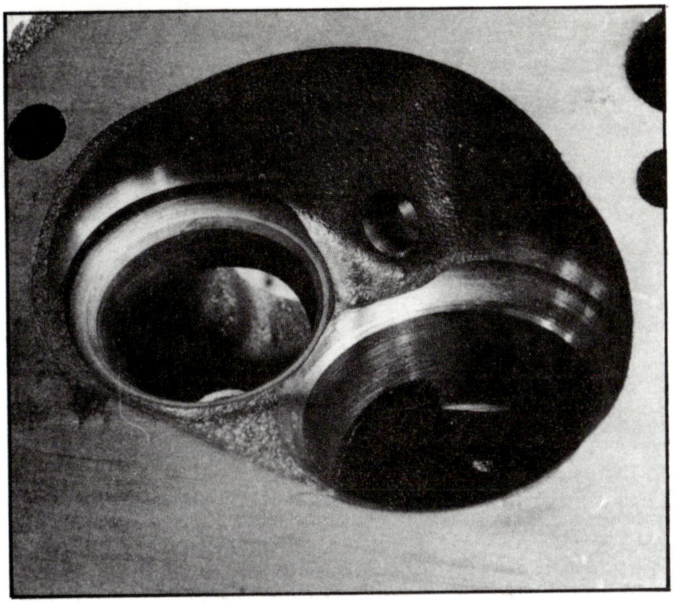

This new Chevrolet cylinder head exhibits a "blued" line around the valve seats, which indicates a factory induction-hardened seat. Designed for unleaded gasoline, such heads are perfect for LP fuel; however, the effectiveness of the original surface hardening will be reduced by subsequent valve work.

HARDENED VALVE SEATS

Hardened valve seats are particularly helpful in a propane engine. Almost any valve seat can be hardened by inserting a magnetic coil inside the valve pocket (after the seat is machined) to heat and quench the iron until the material becomes very hard and durable. But this is not a practical "backyard building tip" for most readers; and very, very few professional machine shops have this capability, though some of the larger factory-authorized engine rebuilders have the necessary equipment. So if the engine does not already have heads with hardened seats, an alternative—though expensive—solution is to replace the heads with a set that already has hardened seats (if such heads are available).

However, it is also important to remember that induction-hardening is a "skin-deep" process. The hardening process extends about 0.030-inch into the head, but only the first 0.006-to 0.010-inch is really hard. Therefore, a cylinder head may lose the majority of the hardening benefits after the valve seats have been recut one or two times. (Actually, one very sloppy valve job can cut through the hardened surface.) Thus it is possible for the valves to recede even if the seats were originally hardened; but in this case, the problem lies in bad workmanship.

Another alternative for older (pre-unleaded) heads and hard-seat heads that have already undergone one or two valve jobs is to install hardened *seat inserts*. Made of special alloys of steel, these seats can withstand numerous valve jobs (though they are tough to grind). Most larger automotive machine shops can install seat inserts, which involves milling away a portion of the valve pockets, pressing in the hardened inserts, and remachining/grinding new valve seats. This work is not cheap; but if durability is a consideration, this investment is worth every dollar.

There are various brands of hard-seat inserts. One that comes highly recommended by propane experts is the Martin-Wells "Well-Tite" brand, which is 42% nickel. Many inserts are made of vary hard *stellite*, but the Well-Tite high-nickel seats are easier to machine and do a better job of transferring heat away from the valves (into the water jacket).

Nonetheless, if you have a choice, heads with built-in induction-hardened seats will transfer heat more efficiently than any type of insert. So, the best setup is to start with a good set of hard-seat heads; and when it's time to do a valve job, have the heads

When working with early heads, or a set of late heads (for unleaded fuel) that have already received one or two valve jobs, special hardened seat inserts are the best answer for a propane engine. A pilot is installed in the new guide to accurately locate the cutter that cuts away the old seat.

The hard-seat manufacturer will specify how much interference fit is required. The new seat is hammered into the head with a driver.

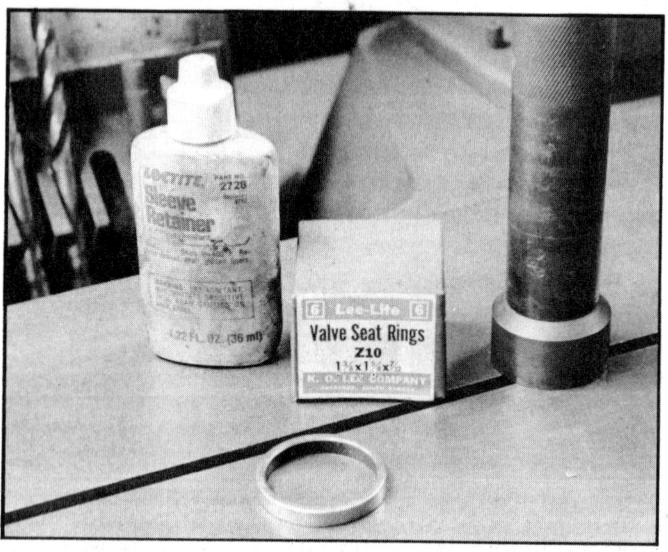

Besides the interference fit, it is recommended that hard seats be installed with a light coat of Loctite-brand "sleeve retainer" on the backside.

machined for hard inserts (if necessary).

VALVE FACE & SEAT PREP

Finding a machine shop that will do precise valve face and seat prep is becoming increasingly difficult. Few readers can hope to accomplish such a job in a backyard garage; but if you realize what is involved in a good job, you'll be better equipped to find a competent local shop. Most machine shops do a reasonable job for an engine that will be operated on leaded gasoline; however, the quality of work needed for a propane engine is usually found only at high-tech specialty shops that can perform the kind of head prep required for high-performance and racing engines. These shops usually have the best equipment and will invest the extra time required to do an exacting job. This extra attention to detail will certainly cost more; nonetheless, valve work is no place to scrimp or cut corners.

Though every machinist favors a specific brand of equipment, most high-tech head shops use this type of oscillating seat grinder.

Whether using hardened seats or not, the flow performance can be improved by using a grinder fitted with an emery drum to blend the valve-seat area into the port throat. It is best to do this before the valve seats are cut—unless you're very, very careful.

After the cylinder heads have been obtained or removed from the engine, they should be *thoroughly* cleaned. If the castings have been used previously, the standard solvent bath or hot-tank cleaning will not be adequate. Heavy accumulations of rust in the water passages are a big problem with old heads, and these cleaning methods won't get all of the rust out of the water jacket. You can take the heads to a rust-stripping shop (where the rust is chemically removed from the metal), or you can give them a do-it-yourself treatment with swimming-pool acid. Mix muriatic acid half-and-half with water, plug the holes in the head, and fill the water jackets with solution. Let it sit for 3 to 4 hours, then drain it off; the inside of the passages will be spotless. However note with care: *This muriatic acid solution is highly toxic*! USE CAUTION; wear a respirator mask approved for toxic chemical vapors, wear rubber gloves, and read and follow all precautions listed on the muriatic acid container.

This technique can also be applied to the water-jacket passages of a cylinder block. This will increase heat transfer and cooling efficiency. But once again, to be absolutely safe, you must note and heed all recommended precautions when handling toxic fluids!

This is a radical case of valve recession in a 440cid Mopar head. These engines are notorious for high temperatures in the combustion chamber. In addition, notice the amount of rust scale in the water passage. Just 0.60-inch (1/16) of scale on a 1.0-inch thick cast iron surface reduces the heat transfer capacity to that of a 4.25-inch thick piece of cast iron!

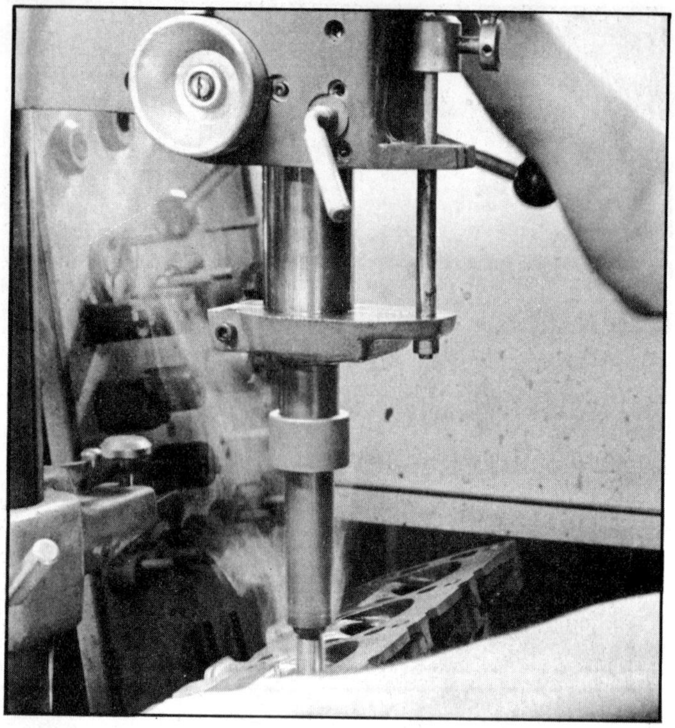

All precision valve work is indexed from the axis of the guide, so guide rework must be the first step during a rebuild. If the guides are worn, they should be bored out and new guide inserts pressed in place. "Knurling" of the valves (to reduce clearance) is not a satisfactory long-term solution to worn or oversize guides.

New replacement valve guides are available in various materials, ranging from cast iron to bronze and special alloys. Here machinist Mario Felix installs new alloy guides with an air hammer.

Since the primary machine tools used to perform a valve job are located from the axis of the valve guides, the first thing that must be shaped up are the guides themselves. Chances are that used heads have some "slop" in the guides; so if you are going to the trouble of rebuilding the engine, make it right from the beginning and *install new guides*. Replacement guides are available in hardened-steel sleeves, aluminum-bronze sleeves, or threaded-coil alloy guides (sometimes called simply "bronze-wall" guides).

If steel or aluminum-bronze sleeves are going to be installed, the existing cast-iron guides are bored out to a larger diameter (completely eliminating the

Shown here with a trio of Hastings-brand replacement guides (an outer shell of stainless steel with an silicon-aluminum-bronze-phosphorus alloy core) are three types of specialty seals: rubber-covered (left), Raymond metal-shell with Teflon inside (center), and Perfect Circle all-Teflon (right). Among stock types, the best in oil control is the rubber "umbrella" used on smallblock Fords.

Replacement guides are generally supplied in unfinished form. After they are pressed in place (lightly oiled with an 0.004-inch press fit), excess material will protrude into the valve pocket (right). This must be cut down from the valve side until the cutter barely exposes the original head material (left). Since the reaming or honing process begins at this end, the bottom of the insert should be chamfered to guide the reamer into place.

stock guides). The replacement sleeves are pressed into the large bores, and then the inside diameter of the sleeve is honed to gain the correct clearance with the valvestems. When Winona brand Bronze-wall guide inserts are selected, the existing guides must be drilled slightly oversize and threaded. Then, the wound-wire inserts are turned into the threaded cast-iron guides, and the inside diameter of the insert is honed to size. Many performance rebuilders prefer the Bronze-wall guides because they offer less frictional contact area and more area for oil retention. And once the guides are prepped for threaded inserts, the inserts can be replaced at a later time without additional machine work (other than honing to size).

When we say "hone to fit," we mean just that. When purchased, new sleeve-type guides should be about 0.003-inch undersize; so they must be honed—not reamed—to gain the proper clearance with the valvestems. If using standard valves, stick to the engine manufacturer's specifications for guide clearance. For valves with sodium-filled stems (aftermarket), you must run a little more clearance—about 0.0035-inch—to allow for the expansion of the stem.

To further ensure a tight, clean-running engine, the tops of the guides should be cut for some sort of "positive-type" aftermarket seals. Any machine shop that can do the work outlined above can easily cut the guides for aftermarket seals. For a top-notch job, run just the aftermarket valve seal on the exhaust valves; but use both the aftermarket seals and the stock seals on the intake valves (producing, in effect, a double seal).

Also, have the machine shop check whether all of the valve-spring pockets in the head are *concentric* to the valve guides. If any of the seats are off-center, the valves may be warped by uneven spring pressure and the guides will wear out prematurely. Most of the time, the concentricity is good; but some engines—like the Ford FE series (352, 390 or 428)—are notorious for this type of problem. Many machine shops don't ever check this, but it is important. If the pocket *isn't* concentric, a new spring pocket must be machined (on center with the guide), and a cup-type shim should be installed to regain the correct installed height for the spring.

If the engine is of older vintage, this may be the best time to install new valves. At the very least, *new exhaust valves should be considered.*

In recent years, some Detroit gasoline and diesel engines have been fitted with heavy-duty valves (for better resistance to higher operating temperatures). If available and applicable, these would be excellent replacements for a rebuilt propane engine. Some aftermarket companies (like TRW, Speed Pro, etc.) produce heavy-duty intake valves (often made of the tougher steel typically used in exhaust valves), and these offer excellent service in a propane engine. Special heavy-duty exhaust valves are also avail-

To reduce oil consumption and hydrocarbon emissions, specialty valve-stem seals are often installed instead of the factory O-ring seals. The top of the guide must be machined to accept these seals. Usually this type of seal is installed only on the intakes, since slight oiling in the exhaust is helpful in reducing exhaust-valve temperature.

After the stem diameters of the valves have been measured, a reamer can be used to rough-cut the inside diameter of the guides to within 0.001-inch of the desired final size. The diameter should then be finished to final size—producing the required stem clearance—with a guide hone.

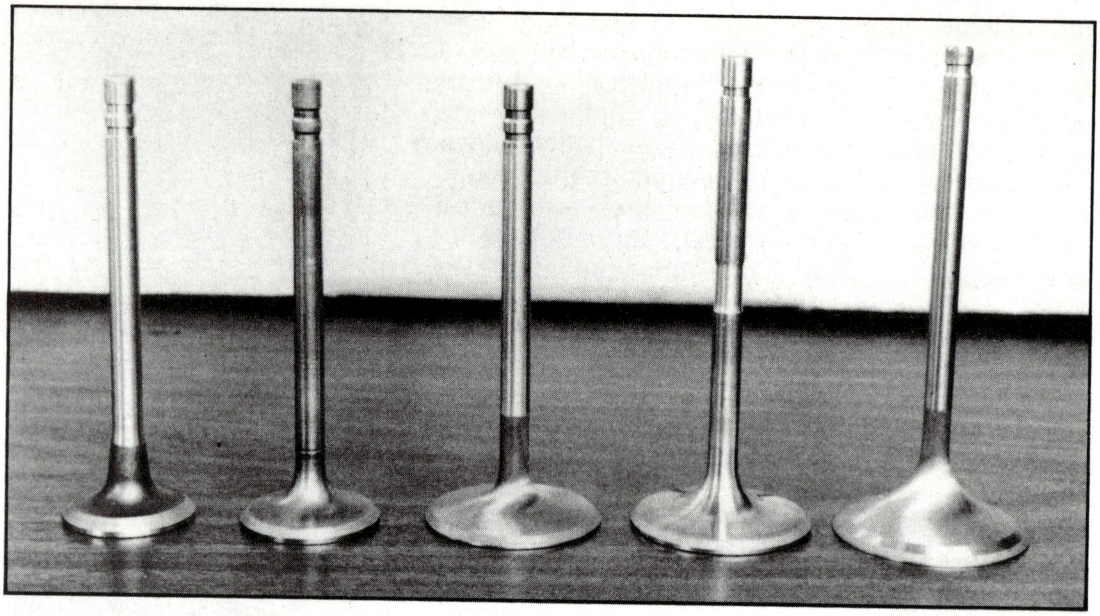

While stock valves are sufficient for most propane engines, the heavy-duty user or performance enthusiast has a host of specialty choices—stainless, titanium, hard-chromed stems, sodium-filled stems, etc. Note the varying shapes behind the head of these valves.

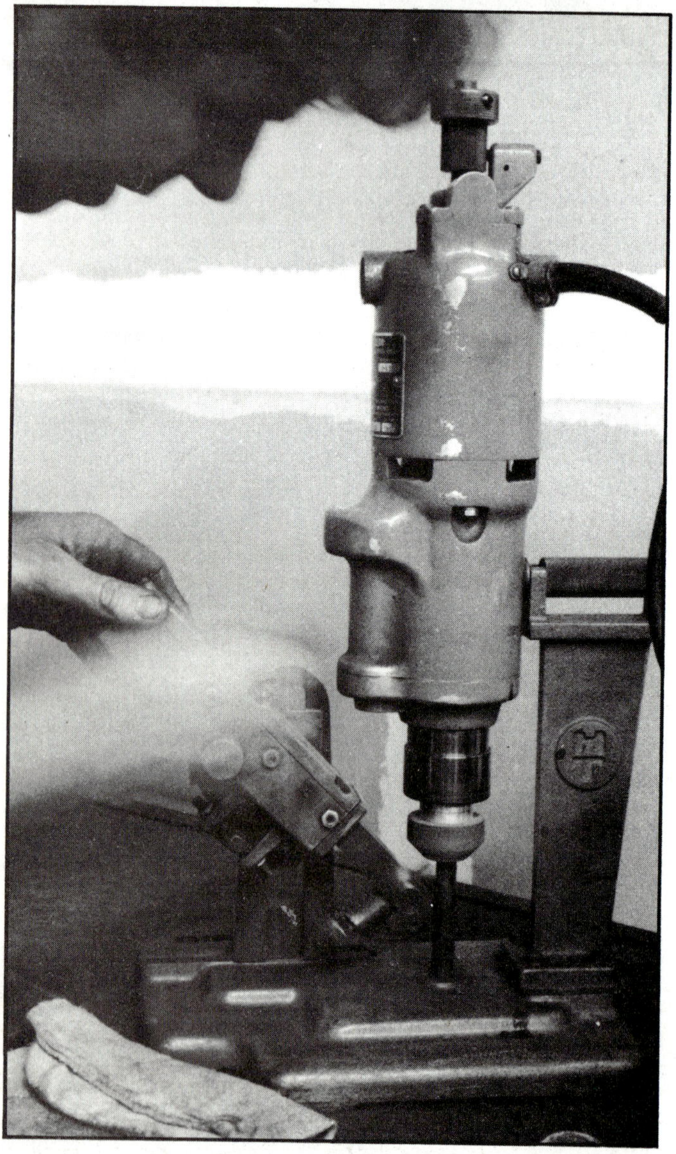

Cutting seats in hardened inserts is tedious work because the stones must be trued several times during the cutting of each seat. This is where the attitude of the machine operator becomes important—he can do either a quick job or a good one.

able for most truck engines (both from OEM manufacturers and aftermarket companies); and besides special alloy construction, these valves sometimes feature hard-chromed stems (to minimize wear) and/or sodium-filled stems (for heat dissipation). If cost is no obstacle, you may want to consider stainless-steel valves. They are very expensive, but they are the ultimate in premium valves. However, most normal-duty propane applications only require a stock or, preferably, some type of heavy-duty replacement valve. (It's generally the seat that fails, not the valve; and hardened seats or hardened inserts should eliminate seat problems).

You may be wondering whether you should use valve *rotators*. Rotators are a good idea for a propane or unleaded gasoline engine because they reduce

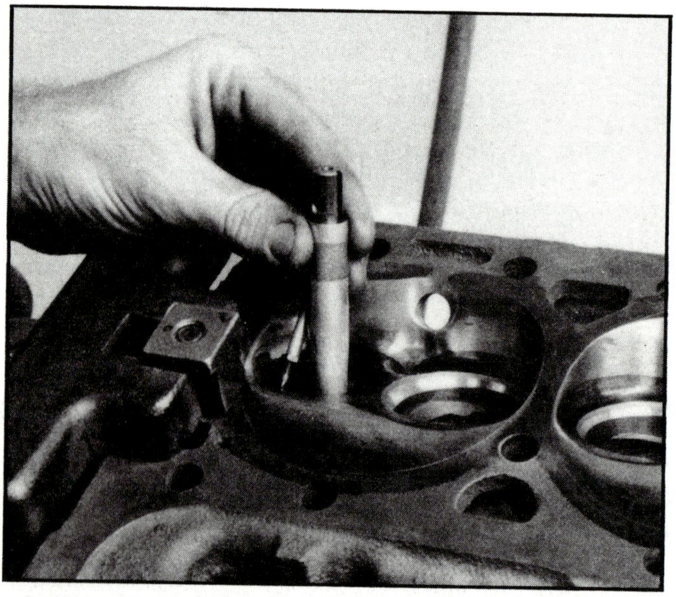

Concentricity of the seat (in relation to the guide axis) is essential! Here the concentricity is checked with machinist's blue dye on the seat and a scriber mounted to a pilot inserted in the valve guide. The seat is concentric if the scribe indicates an even line around the full circumference of the seat.

the welding process and keep the seat clean of materials that could cause a leak (leading to a burned valve). Also, one side of a valve always runs hotter than the other side; so the rotator spreads the heat load around the valve head and seat. Unfortunately, this principle is only as good as the valve springs. When the valve springs weaken, the rotator may allow excessive rotation and the valve may still be turning when it contacts the seat. This will cause the valve to *skid* onto the seat and accelerate seat/face wear. To prevent this, new valve springs should be installed during the rebuild (especially if the valve rotators are retained) and checked on a tension machine to absolutely ensure proper tension at the installed height.

After the rest of the cylinder-head detail work is completed, traditional grinding methods should be used to set the valve seats and faces. The best propane valve job includes: *concentric seats*; standard 45° face angles, *without* an interference angle; relatively wide seats; a relatively short margin width above the valve faces; and even valve contact all around the seat.

The machinist to whom you entrust this work should have top-notch grinding equipment (well maintained) and all the tools necessary to check for seat concentricity. If hard-seat inserts have been installed, the valve job will be time-consuming (because the hard seats wear the grinding stones quickly and the stones need to be re-cut several times during the grinding process). Therefore, many machinists don't like to work on hard seats. And under these circumstances, there is a strong tendency for a lackluster machinist to perform a "quickie" valve job. This usually results in a thin seat with an excessively large margin protruding into the chamber (where it can become an overheated hot spot). Your machinist should be conscientious enough to check his grinding equipment to make sure it trues the valve to the same angle as the stone cuts the seat. One degree of difference in the tools could produce a contact area that is much too thin to promote proper heat transfer from the valve head to the seat.

Valves should also be checked for concentricity. As the stem is rotated, the dial indicator of this K-Line tool checks to make certain the head or stem isn't bent.

The valve seats for a propane engine need to be wider than the typical gasoline valve job: about 1/16-inch on light-duty engines and 3/32-inch for heavy-duty applications. And once again, there should *not* be an interference angle (a standard practice with conventional valve jobs, in which the valve face is cut

Hardened inserts have been installed only on the exhaust seats of this head, where the recession problem is greatest. A good high-performance valve job requires three or four stones to produce a true "three-angle" valve job.

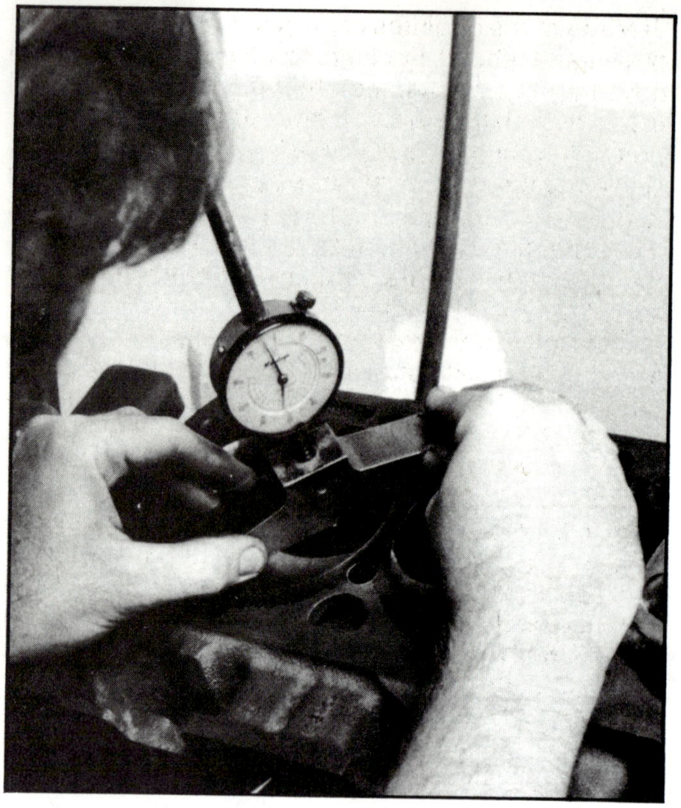

Some meticulous racing shops prefer all the valves to sit the same depth in the head. In the well-known Manning and Brady cylinder head shop (shown here), a dial indicator is used to check valve depth at the center of the valve. To avoid overheating with propane fuel, the valves should extend into the chamber as little as possible.

On this Ford head by Mario Felix the 15° top cut blends the hardened seat insert into the chamber. Note the gray lines on the valves after the lapping test.

to a 45° angle and the seat is cut to a 44°).

Even if the machinist has years of experience, he may not be aware of the seat/face requirements of a propane-fueled engine. Discuss these requirements in advance; and after the job is finished, check the work. This will not require any special tools. With no valve springs in place, seat the valve firmly on the seat (hold it in place with your fingers), shine a flashlight around the perimeter of the valve, and look down into the port for any signs of light leaks. If there are no leaks, push the valve loose (from the stem end); if you feel any friction or "pop" as the valve releases, the valve job isn't concentric. Any misalignment between the guide and valve seat will cause the valve head to stick on the seat. Check this further by coating the seat with machinist's blue dye, then dab some fine valve-grinding compound (or some Bon-Ami cleanser mixed with light oil) onto the valve and rotate the valve on the seat with your fingers. The resultant clean spot in the bluing will be the actual contact patch, which you can inspect for proper seat width and even contact around the circumference. This is, in fact, how a good machinist checks his work while doing valve and seat grinding.

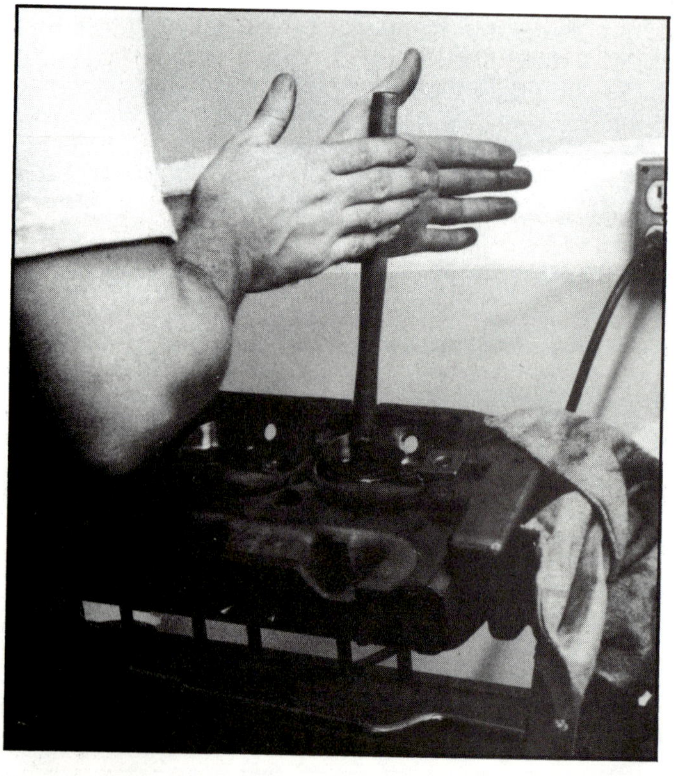

Hand-lapping is no longer considered a practical means of seating valves, but it is a good method for checking the quality of a valve job. Paint the seats and valve faces with machinist's blue, let it dry, and lightly lap the valves with grinding compound. When the valve is removed, the contact area is clearly indicated, and the concentricity and seat width can be measured.

PISTONS, RINGS AND CYLINDERWALLS

Here's an area where you can either take ad-

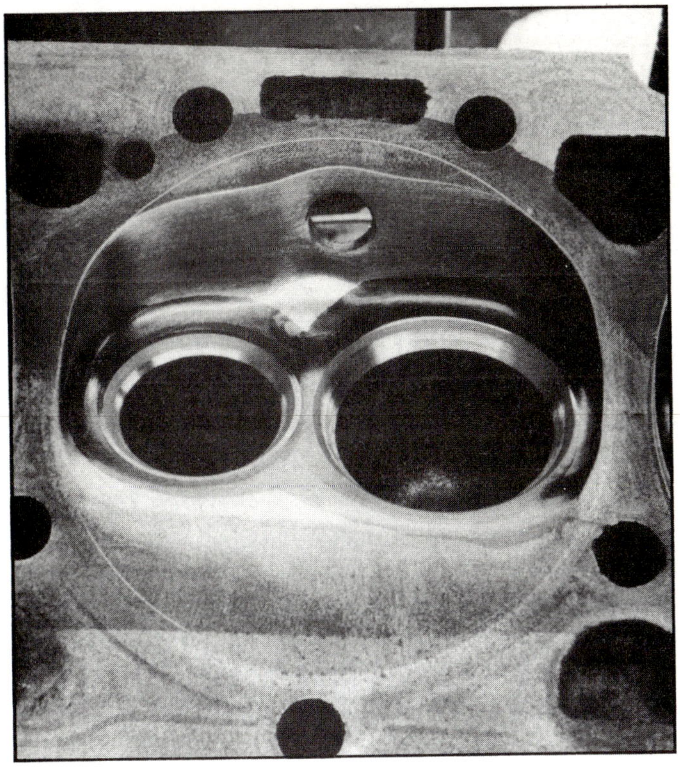

A precision valve job by Manning and Brady (of Reseda, California), after lapping, shows a dull gray line that represents the actual valve/seat contact. Notice that it is even around the entire circumference.

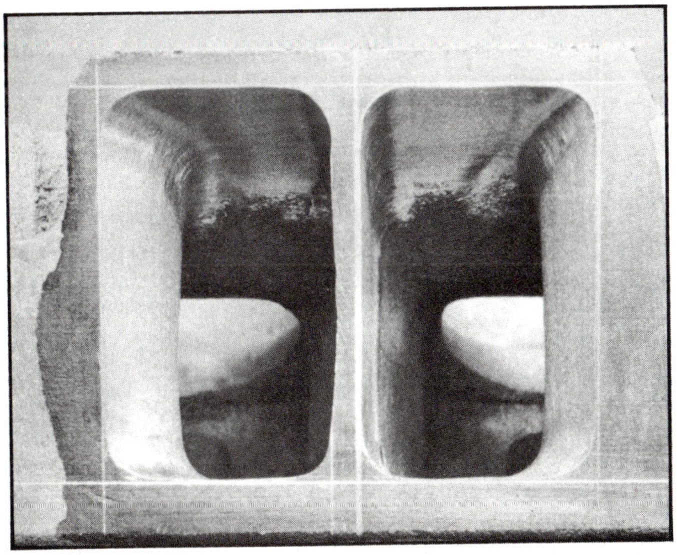

vantage of the qualities of propane motor fuel or ruin the engine rebuild. For instance, it is possible to take advantage of the high octane rating of propane and install new pistons with a higher compression ratio—producing more specific fuel economy and power. Given the ignition-advance requirements discussed earlier, you can usually raise the compression on most engines by as much as two full points (although the service use of the engine will greatly affect the actual maximum usable ratio).

For light-duty passenger-car applications, ratios as high as 10.5:1 can be used without incurring detonation. For light-duty pickups that see only occasional hard work, 10:1 is a good ratio. For heavy-duty or high-rpm use, stay down around 9.5:1 or 9:1 compression.

It will probably be necessary to use aftermarket pistons to get higher-than-stock ratios, but companies like TRW and Speed Pro have excellent pistons (generally available in a wide variety of sizes and ratios) that are only slightly more expensive than stock-type pistons. The new pistons should be fitted to a middle-of-the-road clearance within the factory specifications; i.e., if the factory clearance range is 0.0014- to 0022-inch, use 0.0018-inch.

No matter if it is brand-spanking new or a 100,000-mile veteran, the block must be sent to a machine shop and the cylinder bores must be accurately bored and honed oversize. This will ensure "straight" bores and gain the proper clearance between the piston skirts and the cylinderwalls. When properly done, the honing process will also produce the best surface texture for a "tight" ring seal.

The block should be delivered to the machine shop before the pistons are purchased. Once the block is checked with a dial-bore gauge, the machinist should be able to recommend how much the block should be bored and which specific oversize pistons will be required. In a typical propane engine, properly installed pistons should last virtually forever (because of the reduced wear from friction). However, in colder climates, accelerated skirt wear may become a problem due to long warmup periods at low idle speeds (during engine idle, less oil is thrown onto the cylinderwalls). It is standard practice for many fleet drivers to start an engine in cold weather

Matching the ports on a street engine is a very effective way to reduce flow restrictions and increase performance. This job can be done in a home garage with a grinder and some emery drums. A high polish isn't necessary; just make certain the port and manifold openings match (using the gaskets as templates).

Another useful trick developed by racing shops is to match the ports (intake and exhaust openings in the heads and manifolds) to the gaskets. Begin by blueing and scribing the gaskets outline onto the existing ports.

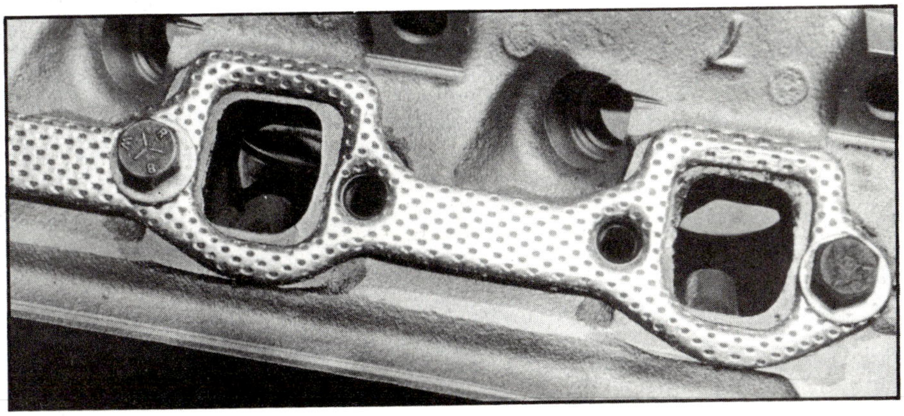

and let it idle while they go back inside for one more cup of coffee (until the cab gets nice and warm). The problem is that propane carburetors don't have a conventional choke and fast-idle system; so in cold-weather operations, propane engines should be fitted with hand-throttles so that the idle speed can be increased during the pre-drive warmup.

Piston rings will also last many times longer with low-carbon propane, but care must be taken when selecting the rings and honing the cylinderwalls. Because the cylinders get excellent lubrication (there isn't a constant bath of gasoline to wash the oil away), the rings don't get overheated in a propane engine and they maintain excellent tension and seal. But the lack of carbon particulate (a byproduct of gasoline combustion) makes it hard for the rings to initially seat or wear-in against the walls. (In a gasoline engine the carbon acts as a light abrasive, and the rings are actually "lapped" to the cylinderwalls after about 1000 miles.)

For this reason, chrome rings should not be used in a propane engine (unless it is a dual-fuel installation, where they can be broken in on gasoline). Chrome rings are used in many production engines for their excellent resistance to abrasion; but if

For intermittent-duty applications, high-compression pistons can be used to achieve excellent performance/economy with high-octane propane fuel.

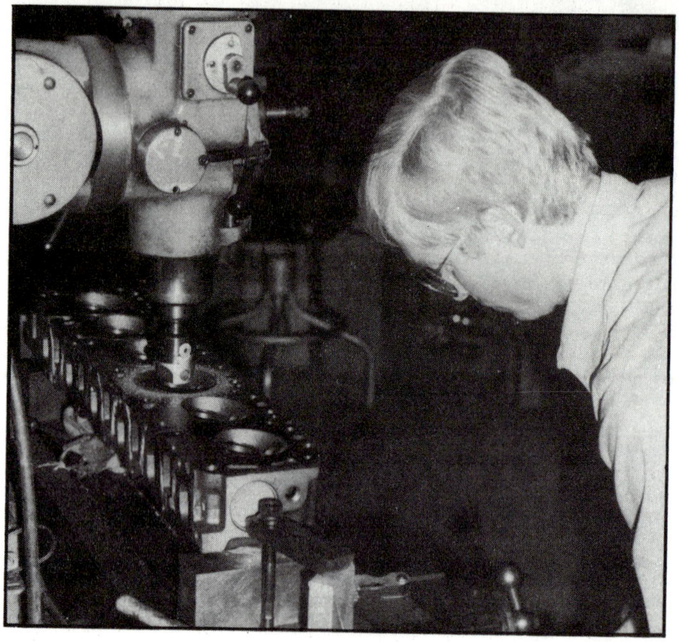

For extra sealing on turbocharged or high-compression engines, the heads should be machined for copper O-rings. An 0.035-inch deep groove is cut around each combustion chamber, and 17-gauge copper wire is tapped into the groove. This wire crushes into the stock head gasket, increasing the seal strength.

Racing-engine builders bolt a thick steel plate to the block to simulate the distortion caused by the head bolts. Using this type of "torque plate" during honing and boring ensures a true bore when the heads are installed and the head bolts are torqued to spec.

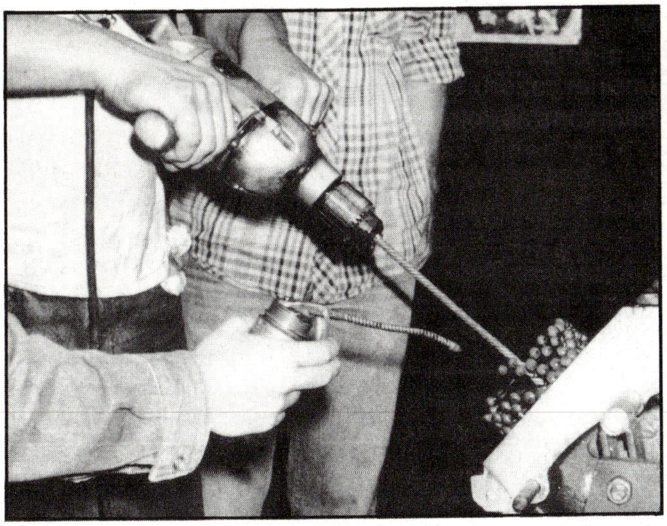

The "Flex-Hone" is noted for producing a cylinderwall finish that retains oil (for ring lubrication) and seats quickly with the rings. Developed by Brush Research Mfg. (Los Angeles, California), the Flex-Hone can be used in a standard electric drill. The nylon bristles are tipped with silicon-carbide impregnated balls and the resulting cross-hatch pattern is very even, regardless of the cylinder taper.

chrome rings are used with propane, the engine should be run for 1000 to 1500 miles on gasoline before the conversion. Otherwise, it could take as much as 50,000 miles for chrome rings to properly break-in on propane; and it is even possible that they may never seal properly—with a resultant increase in oil consumption.

The best type of rings for a propane engine are the traditional plain cast-iron compression rings (with high-temperature molybdenum face coating, if you prefer this type of ring), a standard cast-iron second ring, and a conventional three-piece oil ring with chrome rails.

Certainly the method of cylinderwall finishing has as much to do with the ring life, seating, and oil consumption as the rings and pistons; and bore finishing is a subject on which every engine builder seems to have his own pet theory concerning what works and why. Just as with valve guides, the cylinder bores should still be about 0.002- to 0.003-inch undersize after the initial boring process is completed. The bore is then brought to the final finish diameter (to gain proper running clearance with the selected pistons) during the honing process.

Most shops prefer to begin rough-honing with 220-grit stones, and finish with 280-grit or 400-grit stones. Cleanliness is important during honing (and every other step of engine rebuilding); and the hone should be cooled and lubricated by special honing oil, which contains the proper ratio of cutting oil, lapping oil, wetting agents, and rust removers. Remember, however, that grit numbers of various hones are only comparable between hones of the same manufacturer; so even though we give these general guidelines, the specific honing techniques and hone numbers used by some shops may differ. An experienced shop will know how to get the best results from their specific type of equipment. If you tell them that the engine will be fueled with propane and what kind of rings you will be using, a good shop will produce a suitable wall finish.

Many performance-engine builders also like to use a Flex-Hone (TM), made by Brush Research Mfg. Co. (4353-4355 East Floral Dr., Los Angeles, CA 90022) to finish the bores. This is an excellent hone that promotes even and quick ring seating. The Flex-Hone is a large brush with balls of silicon-carbide on the ends of the Nylon bristles; it provides a gentle, even pressure on the cylinderwalls during honing, and it produces an excellent cross-hatch pattern.

High-performance engine builders have discovered some tricks that can be beneficial for down-to-earth mechanics. (Obviously, if a NASCAR engine builder can make a 600-horsepower engine live through 500 miles of high-speed racing, he's learned a few good tricks!) One racing-developed engine-building trick that's no longer a secret is the use of a *torque plate* while boring and honing the block. When a cylinder head is bolted down to a block, the bores are distorted by the stress imparted to the block by the heads bolts. So the sharp machinists have special plates made up of thick steel that roughly duplicate the dimensions of a head gasket. They bolt this to the block and torque it down (just like it was a cylinder head with open access to the bore) so that the block can be in its "normal condition of distortion" while it is bored and honed. (To carry the torque plate idea a little further, it also makes sense to have the torque plate bolted in place on the cylinder head when the valve-grinding work is being done because there may be some distortion of the head as well as the block.) Torque plates for many domestic engines can be purchased through speed-equipment shops and mail-order warehouses.

Regardless of the hone type used, the engine must be thoroughly scrubbed with hot, soapy water and a non-metallic bristle brush after the machine work. Keep scrubbing the cylinderwalls until you can pass a clean rag through the bore without finding any trace of dirt. Immediately after the cylinderwalls are cleaned, they *must be oiled down to prevent surface rust*, which will begin to form even before the walls are dry.

Speaking of oil, what we said earlier about changing oil when you "lose your nerve" may be a little misleading. Oils break down with continued application of heat. Therefore, how hard the engine is working and how hot the oil is running will have a telling effect on the oil. In a rough-duty application, the oil is going to turn into jelly pretty quickly; but since every application has different heat and load characteristics, it's hard to predict exact oil-change intervals. Special oils *are* made strictly for gaseous-fuel engines, but these are recommended primarily for stationary industrial engines and aren't necessary for cars and trucks.

CHAPTER 3

PROPANE INSTALLATION

PROPANE INSTALLATION

In many aspects of life, a *little* knowledge can be dangerous; but when undertaking a propane conversion, the lack of thorough knowledge can be more hazardous than usual. The inevitable results of misinformation and sloppy workmanship (especially in safety-related aspects) have caused the established propane industry to concentrate on supervised fleet conversions and, for the most part, ignore the average do-it-yourselfer. However, we have to assume that a few not-to-be-deterred individuals will attempt home-brew installations without proper knowledge or experience (if they can get the equipment); and though we don't want to encourage privateers, we do want to point out the important technical aspects of a professional propane-fuel conversion.

The knowledgeable reader who plans to have a conversion completed by an authorized service center will benefit from this overview. It's important to know the *correct* way to do the job. This will help the owner to intelligently evaluate the work of the conversion specialist, and it will hopefully ensure that the owner understands the importance of proper operating and maintenance procedures.

Given the amount of advice presented in the introductory chapters, you might expect a conversion to be a complicated effort that requires several days. On the contrary; an experienced installer can handle a typical pickup conversion—like the one illustrated here—in a single day. A more complex passenger-car installation might take, perhaps, a day and a half (barring any major unexpected stumbling blocks). More than anything, this highlights the simple and straightforward nature of propane fuel systems; but it must be remembered that professional installers have a lot of experience and the proper tools/accessories to complete the job efficiently.

Most of the general technical information contained in the first three chapters "must be done" to complete a safe installation; however, some of the specific techniques are really of the "should be done" variety. When a private individual brings his (or her) car or truck to a conversion center, a competent mechanic will do all of the former, but a really top-flight mechanic will also do most of the latter. It is important to know which is which.

This is not to indict the many fine, experienced mechanics who are performing quality installations, but we have to remember the current nature of the business. Because of the fluctuations of interest in alternative fuels, the conversion business sometimes becomes overrun with enthusiast customers; as a result, price competition becomes quite fierce. Unfortunately, the average motorist usually doesn't know the first thing about how the job should be done; so his decision is often based strictly on *price*, with little consideration for *professional quality*.

The customer who is shopping by price alone doesn't know that the shop advertising a $999 special in the Sunday newspaper doesn't do as much work as the shop that charges $500 more. Unfortunately, to give a competitive price estimate (and still make a decent living), an installer is often forced to limit how much time and effort he can invest in each job. Every performance tip that we advised in the engine-building chapter is helpful (though it can be expensive when you consider that the average shop labor rate is $20/hour or more), and the simple economic truth is that the conversion center can't perform all of the "should be done" work and still keep their price competitive. However, it is safe to say that, like anything else, you generally get what you pay for.

Typically, a bare-bones installation will include the following work: (1) mounting the propane tank, (2) fabricating and installing the high-pressure lines, (3) venting the tank pop-off valve, (4) plumbing the filler and outage valves outside the body (passenger car only), (5) removing the stock fuel pump and blocking off the stock mount, (6) removing the stock gasoline carburetor and replacing it with the propane mixer, (7) adapting the carburetor linkages and automatic transmission kickdown to fit the mixer, (8) mounting and plumbing the vacuum, vapor, water and fuel lines to the converter, (9) installing the filter/fuel-lock, and (10) modifying the advance curve of the ignition distributor. If it's a good shop, they'll also run the vehicle on a chassis dyno, before and after the conversion, to compare performance and properly set the power-enrichment adjustment on the propane mixer (this job should only be done when the engine is under load). What you drive away is a vehicle that runs on propane.

For that vehicle to run *well* on propane—to take advantage of the fuel's benefits and avoid its shortcomings—more work needs to be done: the "should be done" efforts. Among the jobs seldom done by a commercial conversion center that fall into this category are: (1) change the thermostat to a 160°F type, (2) upgrade the ignitions system (if it isn't a new vehicle with factory high-energy ignition and silicone wires and boots), (3) re-gap the sparkplugs to .030-inch or install new properly gapped plugs, (4) remove the intake manifold and install gaskets that will block-off the heat-riser passages, (5) perform a precision valve job, (6) install hard valve-seat inserts (if appropriate), (6) duct the air-cleaner snout to pick up cold air, and (7) optimize the compression ratio (if appropriate). Unfortunately, some of these tasks are quite time consuming; and if they were required in every instance, they would greatly increase the price of the average propane conversion!

In most cases, the best bet for the average owner is to shop around *carefully* before selecting a conversion center. Of course, cost is important; so if the price seems reasonable, find out exactly how much work they do for the quoted price. Spend a few minutes browsing around the shop and look at the general condition of the equipment (you may not know anything about machine tools; but if the equipment is clean and if the shop is orderly, chances are good that

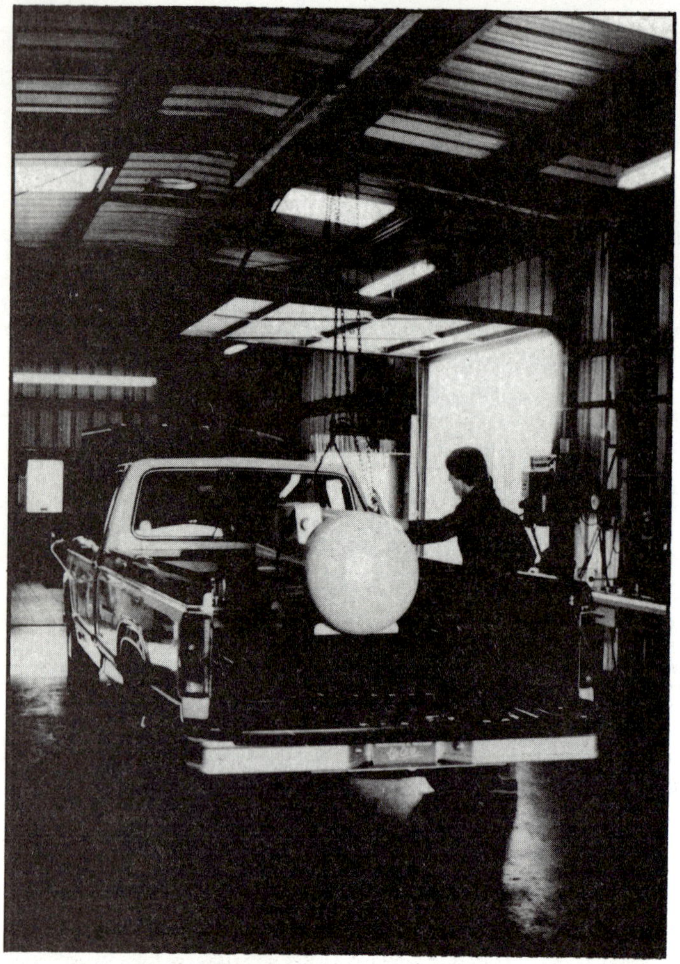

The first step with most propane conversions is the mounting of the LPG tank. In this installation, Gene Carr (Mutual Propane, Tustin, California) used an overhead crane to position the tank in the truck bed. The tank is painted white to reflect heat.

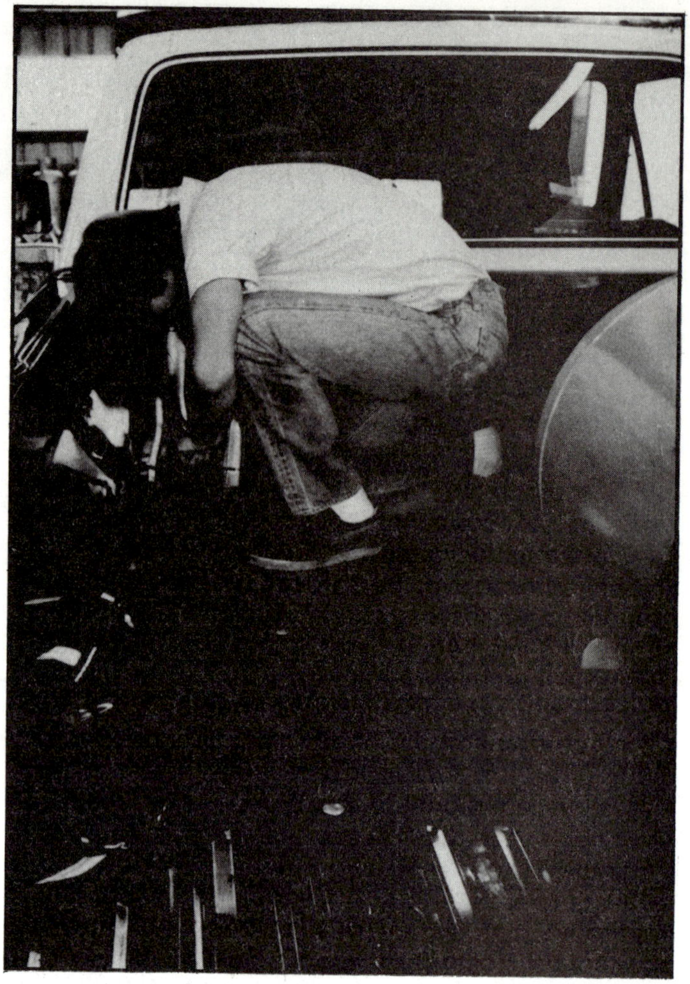

With the tank in position, the bed is marked, making certain the mounting points are not located over a brake line or other critical chassis component. The tank is then removed and the mounting holes drilled.

the installer takes pride in his work). Find out how much experience they have, and get some references to previous customers—and call these customers to see what they have to say about the shop (most people will be glad to give a brief appraisal, and they can often provide some helpful advice).

Once you locate a suitable installation shop, you can turn the job over completely to the professionals; but if you have some mechanical ability and experience, you may want to undertake some of the collateral work yourself. Much of the "should be done" work in a typical installation can be performed rather easily with the proper tools and a factory rebuild manual (even if valve-seat inserts or a valve job are required, you can save money by removing the heads, delivering them to a machine shop, and reinstalling the refinished heads); this will be considerably less costly—provided you are reasonably proficient—than having the entire job finished at the installation shop.

Older vehicles will probably require valve-seat inserts and valve work (if not built for unleaded fuel), but new vehicles present entirely different problems. While they have hardened seats and good ignition systems (making them highly suitable for propane conversions) newer vehicles also have an endless array of hoses, vacuum lines, switches, and valves that complicate repair work. It's going to be up to you to determine the final disposition of these components. Depending on the year of the vehicle and the emissions laws of the particular state where the vehicle is registered, you may be able to remove some or all of this equipment as part of a propane conversion. On the other hand, depending on state laws, *the conversion shop may not be legally allowed to change anything connected with the factory emissions equipment*, including the ignition-advance curve.

This is a tricky situation. The conversion mechanic may know he isn't *supposed* to alter anything; yet in some cases, he may also know that the factory advance curve isn't suitable for optimum propane performance (and in some cases, engine damage could even result). Luckily, the advance curve isn't something that is obvious to anyone looking at the engine (though it can be checked with electronic testing equipment). Sometimes, however, what the law says

The tank is bolted down. Silicone sealant is used around the washers to prevent rust formation in the drilled holes. If practicable, the mounting holes should be located through a frame member or some similar strong point. If the tank must be bolted to thin sheetmetal, reinforcing plates should be used on the backside.

is illegal for a garage mechanic (with a stiff fine as an added deterrent) may be only lightly discouraged or completely ignored when the individual car owner does it himself!

In the final analysis, each owner must make some important decisions before undertaking a propane conversion. Local laws should be clarified (a good mechanic is often the only source of information about laws that pertain in a specific area, but this is certainly not foolproof), and it is usually a good idea to check a service manual carefully to determine what emissions components are present and how they work. This information will help the owner and the installer determine what equipment to retain (with the new propane equipment) and still ensure that the engine runs efficiently.

When a well-equipped professional installation center is contracted to perform the conversion (and if you have carefully completed the the pre-installation preparation) the finished job should provide years of trouble-free, economical service while taking full advantage of the unique nature of propane as a motor fuel.

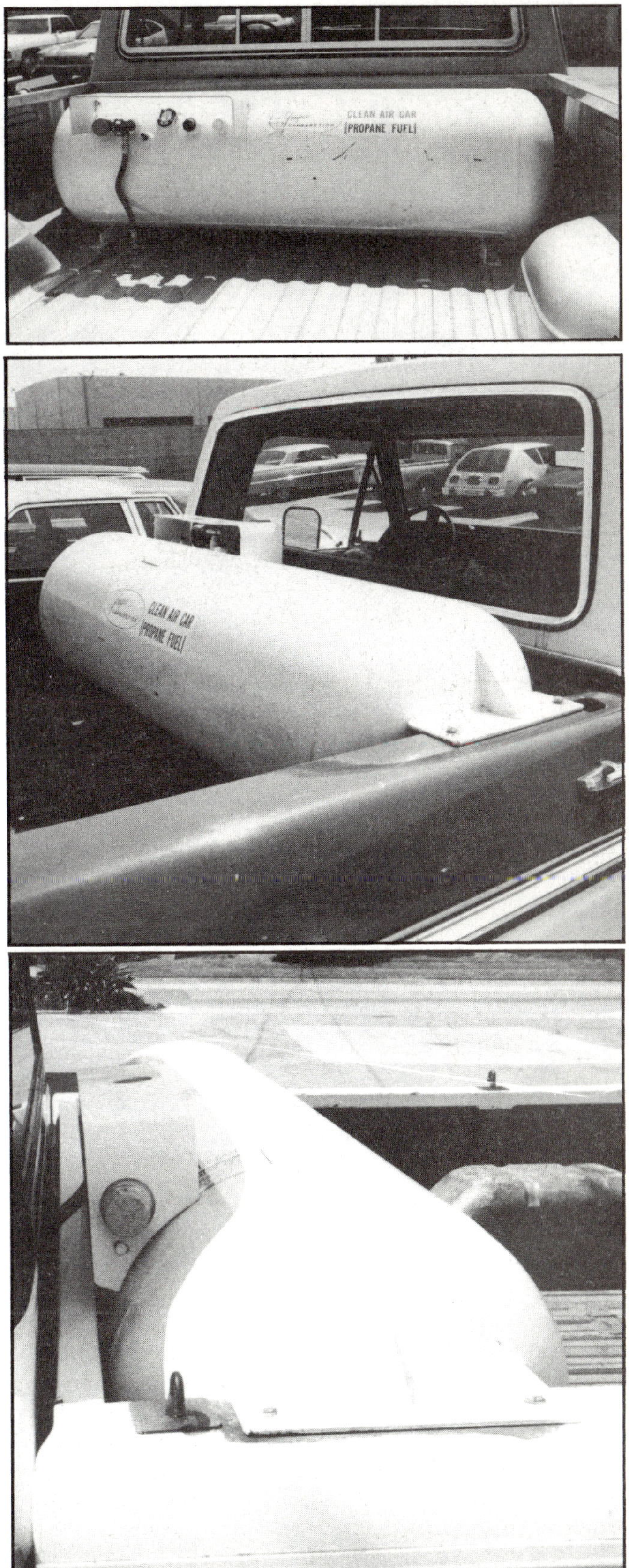

There are several types of tanks available for pickup trucks: (A) most have brackets for mounting to the floor—shown here with a rear-facing filler valve; (B) some have brackets for mounting to the bed sides—these are available for only certain makes and models; (C) some also have a side-facing filler valve—easier to fill when the truck bed is loaded with cargo.

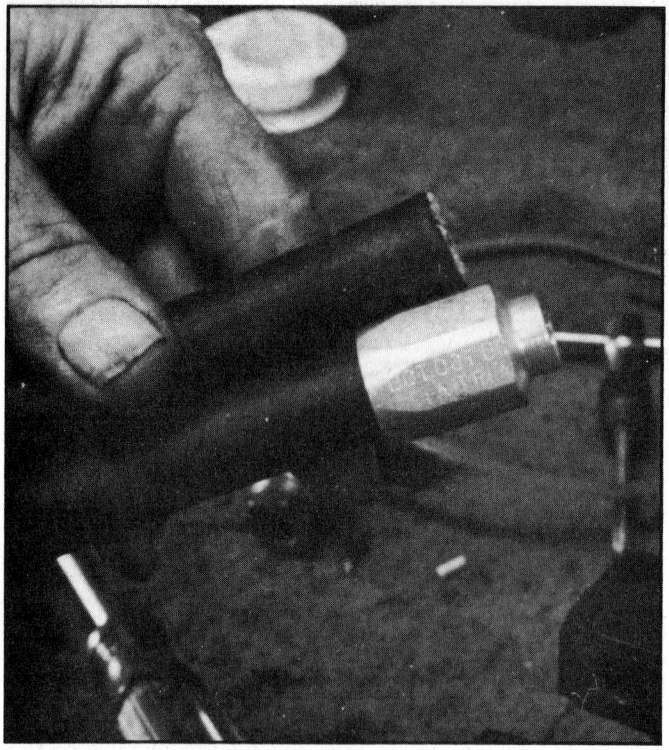

The next step is to plumb the supply line forward to the engine. Assembly of the high-pressure liquid lines is a critical procedure. This should be done by a mechanic experienced with high-pressure hose assemblies. Two-piece (threaded) hose-ends should be marked DOT-approved (Department of Transportation). First, the braided-steel reinforced line is cut to length (before it is cut, the hose is wrapped with masking tape to reduce fraying), and the fitting is pushed on as far as it will go.

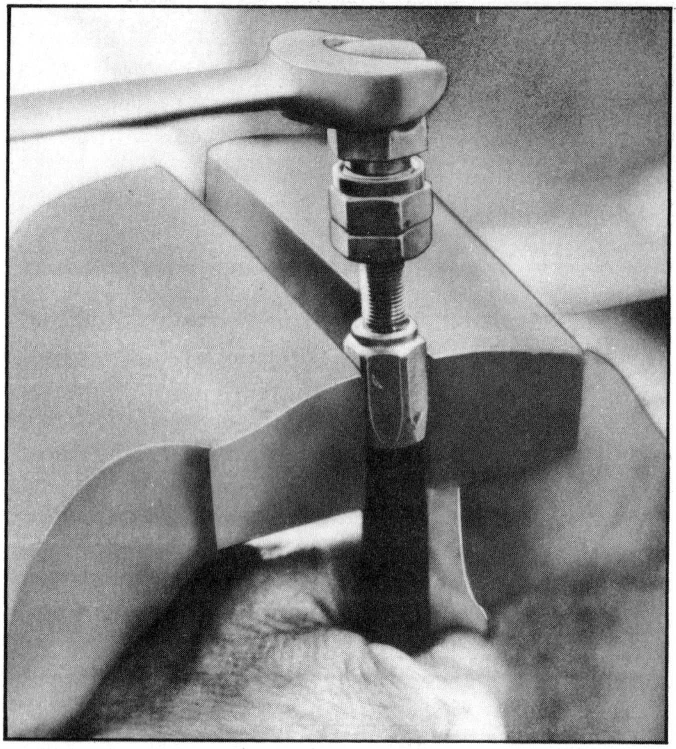

The locking collar of the fitting is lubed with oil and turned with a wrench until it bottoms completely. While the locking collar is being installed, the hose should be pushed firmly into the fitting so it will seat fully over the locking grooves.

A properly assembled hose end is able to withstand all pressure requirements of a typical LPG system.

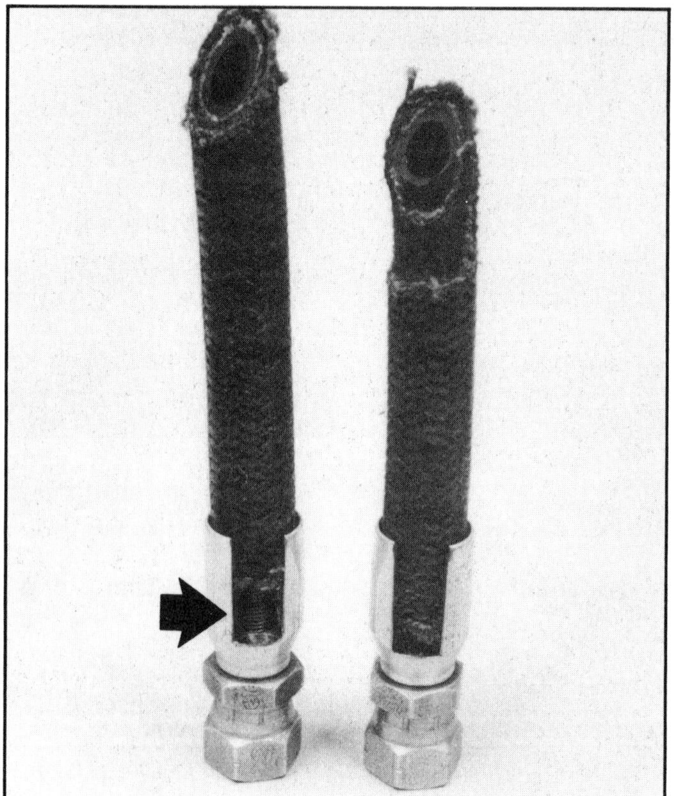

This cutaway of two hose fittings illustrates why pressure must be kept on the hose as the fitting is being assembled. Although both of these fittings look safe, the hose on the left is not fully in contact with the locking grooves. If the hose is not properly located on the locking grooves, the assembly could pull apart.

LPG equipment manufacturers are always concerned about inexperienced mechanics and backyard "improvisations." Simple mistakes, like a badly assembled hose fitting, are potentially dangerous. In this case, a "hack" mechanic didn't have the correct fitting, so he cobbled-up this haphazard combination.

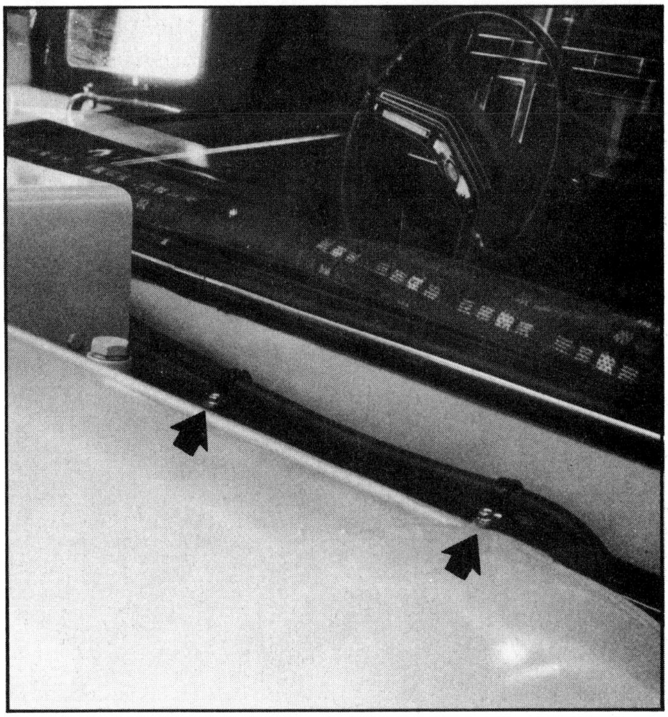

The high-pressure line is threaded to the main valve on the tank, routed forward under the chassis to the engine compartment, and secured every 18 inches (arrows) with clips. On pickup installations, the hose must not be routed where cab/bed movement can cause chafing.

The stock 1-barrel carb has been removed from this 300cid 6-cylinder. Notes are kept to indicate the vacuum source of each line—either ported or manifold.

Gene selected an Impco carburetor base that matched the pattern of the stock carb. He drilled the throttle arm to accept a ball-stud that would snap into the stock Ford throttle linkage (including the factory cruise control).

Manufacturers of propane carburetors offer a wide variety of carburetor bases and adapters to fit virtually any engine. In some cases, a stock gasoline-carburetor throttle base bolts directly to the propane mixer.

The Impco mixer attached to the base is as compact as the stock carburetor. For extra safety, Gene has added a second throttle arm to the shaft and drilled it for an additional return spring. Arrow indicates the idle-mixture screw.

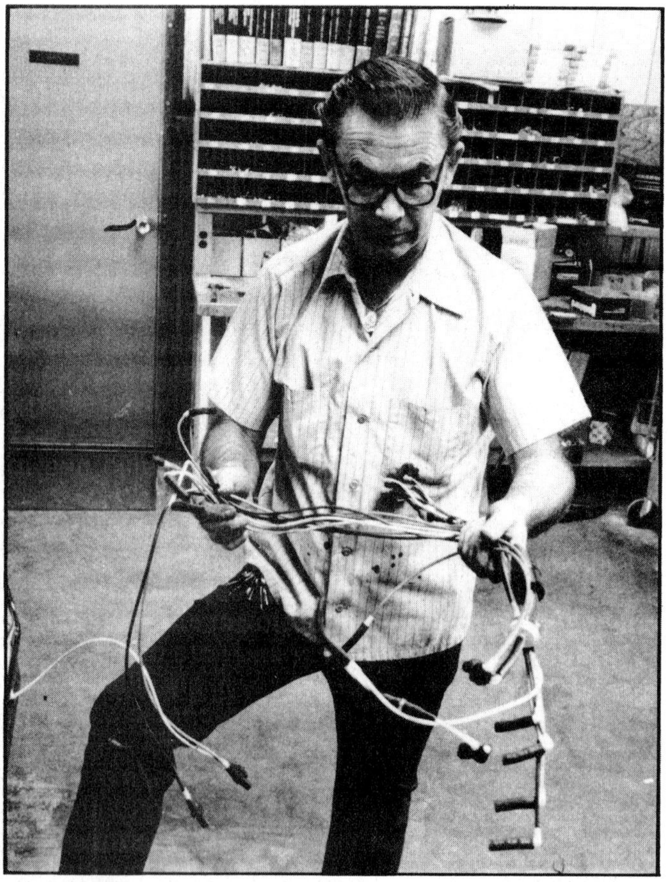

Since this truck is for an out-of-state fleet (not subject to stringent California regulations for '78 and later vehicles), Gene was able to remove the frightening maze of vacuum lines and other appendages in the factory emission package. This truck will run cleaner and far, far longer on propane than it ever could on gasoline.

This vacuum canister on the right fenderwell had to be relocated rearward to provide room for the propane vaporizer and the filter/fuel-lock.

A sheetmetal bracket, designed to hold both fuel-lock and vaporizer was mounted to the fenderpanel, and the fuel-lock was bolted in place. The inlet fitting (arrow) is pointed to the rear of the engine compartment to receive the main fuel line leading forward from the bed-mounted tank.

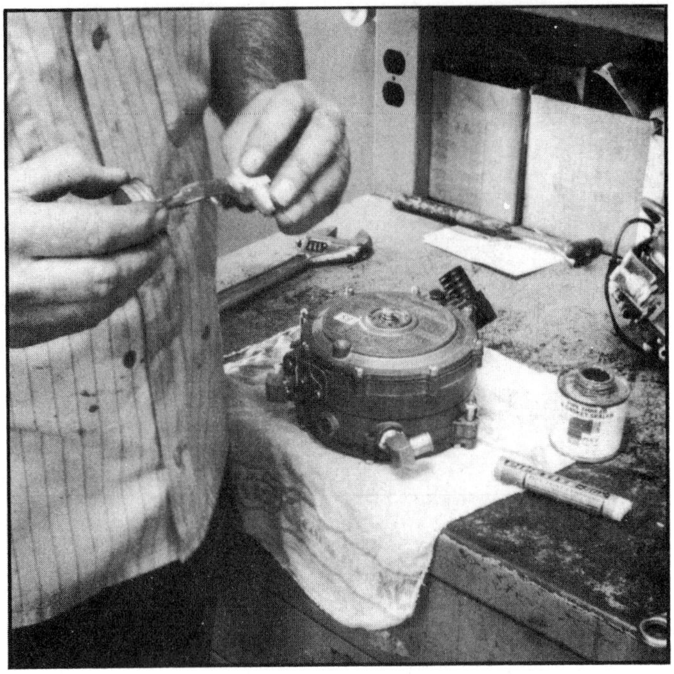

Gene uses two types of sealant on the fittings threaded to the Impco vaporizer: (1) a standard liquid pipe-joint compound (plumber's type) on the propane fittings and (2) a dry, white type (Pipetite-Stik brand) on the two brass water fittings. Never use steel water fittings, because electrolysis (between the dissimilar metals) will corrode the vaporizer casting.

The sheetmetal mounting bracket for the vaporizer and the fuel-lock is designed to slant the vaporizer toward the vapor-outlet fitting. This keeps oil that precipitates out of the vaporized propane from accumulating on the diaphragm.

Gene has pinched the stock heater hoses with clamping pliers (to minimize antifreeze loss) and the two hoses are cut near the firewall in preparation for plumbing the vaporizer.

Both "H" and "Y" styles of heater-hose interrupters are available. You can't simply interrupt one heater-hose line and plumb it in and out of the vaporizer. These fittings allow for proper water circulation even when the heater is not in operation.

The white sealant (Pipetite-Stik) is used on the vapor-hose fittings, which screw into the vaporizer and the Impco mixer. Be careful not to tighten these pipe fittings too much or the metal castings may crack. Teflon sealant is not recommended because it encourages excessive tightening and promotes cracks.

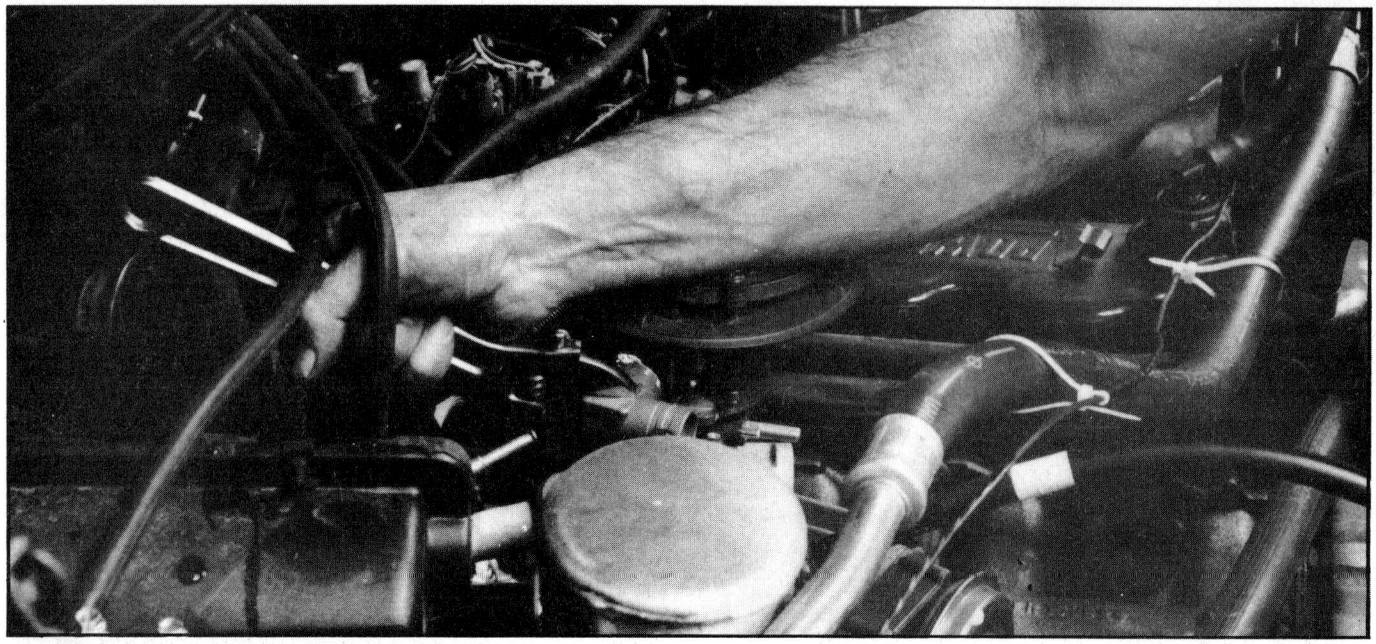

Here the water hoses are hooked to the vaporizer with the "H" connector in the background (arrow). A short high-pressure line connects the liquid outlet of the filter/fuel-lock to the liquid inlet of the vaporizer.

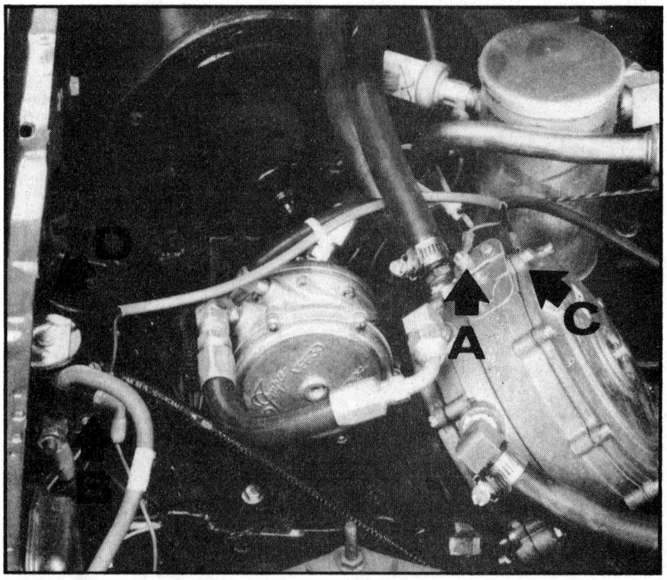

Hookup of the electric primer on the Impco vaporizer is simple. The "plus" post on the vaporizer (A) is connected to a source of start-only voltage (terminal "S") on the Ford starter solenoid (B). A ground wire from the vaporizer (C) runs to a ground on the steel fender (D). The latter wire would not be necessary in most instances, but in this case, the inner fenderpanel of the truck is plastic.

Gene has removed the stock mechanical fuel pump from the block and is installing a blockoff plate.

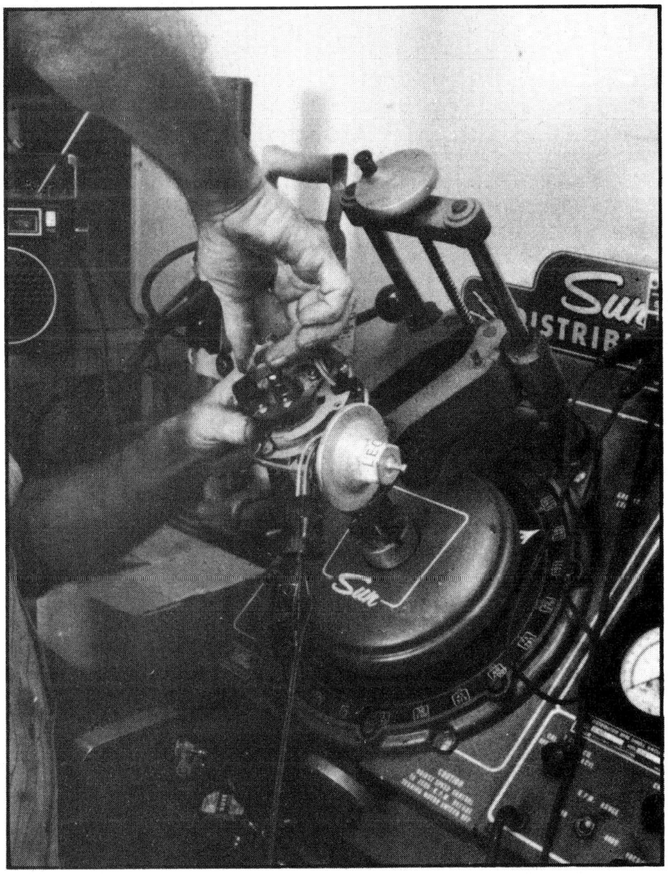

When the Ford pickup was initially tested on the chassis dyno, the stock smog distributor showed only four degrees advance at 3000rpm. In stock form, the late-model distributor was fitted with heavy centrifugal-advance springs to produce a relatively slow rate of mechanical advance. Here, Gene removes the electronic ignition pickup and reluctor for access to the advance weights.

Fuel-pump blockoff plates can be found in most speed shops or you can make your own from flat steel or aluminum plate (using a stock fuel-pump gasket as a template).

Ford distributors have two slots (of different lengths) in the advance mechanism. In this case, one slot allowed for 21° and the other provided 16° of advance. Gene used the 16° slot and brazed it narrower to allow only 10° of advance. These 10° of advance in the distributor furnish 20° of crank timing (remember, the crank turns twice for every one revolution of the distributor). With initial timing set at 10°, this provides a total advance of 30°, just about perfect for a propane engine.

Even if the stock distributor provides the correct amount of centrifugal advance (highly unlikely), it is important to disassemble, clean and lubricate the mechanical-advance mechanism (use special distributor-cam grease). The advance mechanisms in many late-model distributors are not adequately lubricated and are prone to malfunction.

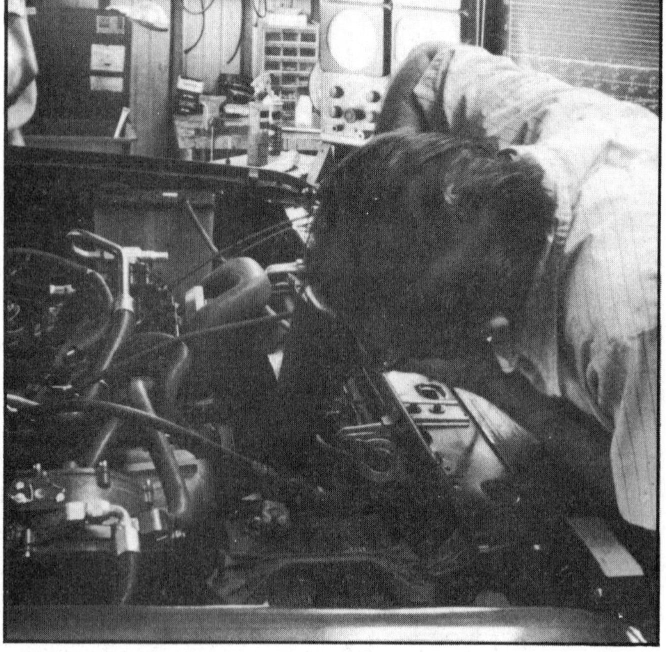

Back on the chassis dyno, Gene rechecks the static and centrifugal advance with Sun test equipment. To properly set the power mixture on the propane mixer, the engine must be operated (under load) on a chassis dyno, and critical adjustment finalized with the aid of a tailpipe-emission analyzer!

With the distributor in place, the engine fired up immediately and idled as smooth as silk. After fire-up, the first priority is to check for leaks. Gene squirts a small amount of soapy water on each fitting and joint (even the smallest leak will cause the soapy solution to bubble). This is important! Don't rely on your sense of smell to locate leaks!

On a V-8 engine or passenger car, the procedures would be the same, except for tank location. This Dodge passenger car is fitted with Impco equipment and an Edelbrock aluminum manifold. With the air cleaner in place, it doesn't look much different from stock, but the engine should provide at least 200,000 miles of reliable service.

This is an improperly installed passenger-car tank. The tank is mounted beneath the body, in place of the stock gasoline tank. In most cases this is not a problem, however, in this installation the filler and outage valves are unprotected in case of a rear-end collision!

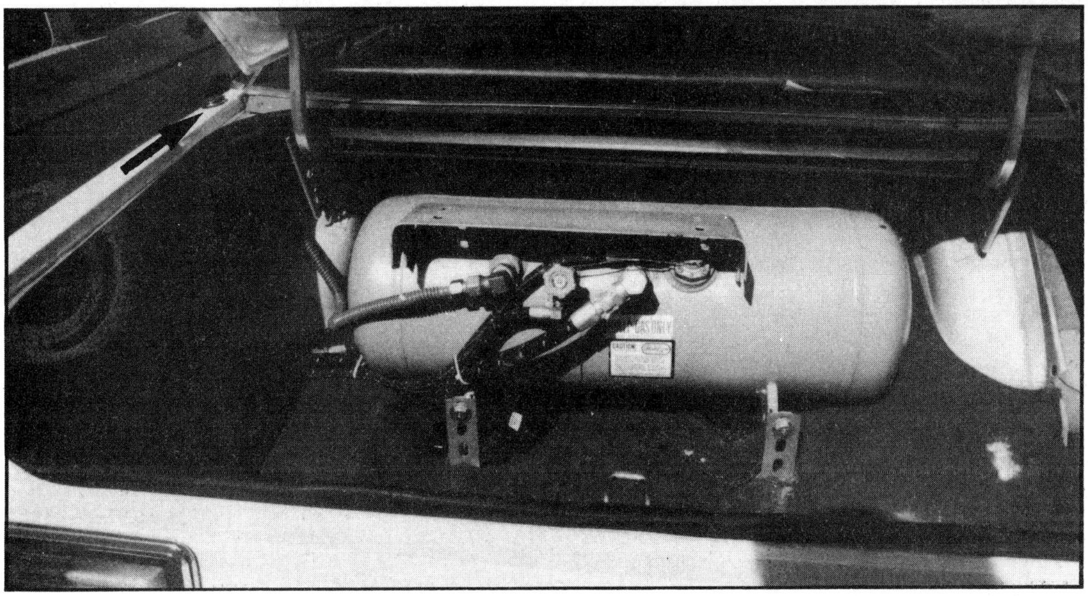

A proper passenger-car conversion is almost always more complex than a truck installation. If the tank is mounted in the trunk, to be legal, the trunk must be completely sealed from the passenger compartment with a metal fire wall. In addition, all valving must be vented to the exterior of the body. In this case, the fill and outage valves are ducted to a strong (heavy-wall steel) exterior box, and the pressure-relief valve is ducted (with flex-line) to an external pop-off valve (arrow) pointing up and away from the body.

This external pop-off for the pressure-relief valve (from Gann Products) disrupts the smooth body contours; but it is essential for a safe, legal job.

This heavy steel box (from Gann Products) has welded fittings for external filler and outage valve vents. A locking cover prevents tampering.

CHAPTER 4

DUAL-FUEL INSTALLATION

DUAL-FUEL INSTALLATION

Given the inobvious nature of most propane refueling stations, a dual-fuel installation—a conversion that allows the engine to operate on either propane or ordinary gasoline—is the most reassuring type of system for the average motorist. The seemingly limited availability of propane fuel appears to be a liability for the long-distance traveler. This limitation is not to be taken lightly, but veteran propane users consistently travel long distances—through all parts of the country—and find that locating suitable propane fueling stations is only a minor inconvenience, at most. Detailed listings of propane retailers are available from most major distributors and rest stops on long trips can usually be planned to coincide with available fuel sources. But still, there are many users who only feel safe if they have a gasoline system as a *backup measure* in case of an emergency.

That word "emergency" is very important. We mentioned before that there are some mechanical drawbacks to a dual-fuel installation, and how the user plans to utilize the two fuels will determine just how serious the drawbacks will be. If the engine will be required to burn either fuel *on a regular basis*, most of the engine modifications that improve propane performance will be impractical. An engine set up for maximum efficiency on gasoline won't produce maximum performance on propane, and vice-versa. If, however, the engine will be operated almost exclusively on propane fuel, and the gasoline system will be strictly viewed as an *emergency backup*, then many of the modifications designed to improve propane performance will still be suitable. The engine may have a tendency to "ping" on gasoline—primarily because of the relatively large amount of ignition advance that a propane-fueled engine can tolerate—but if the engine is not strained excessively (the driver avoids heavy loads, hard acceleration or long hill climbs) it'll get to the next propane dealer with no trouble. (However, the Autotronic Controls dual-curve ignition module described earlier will readily offset this problem; you can set up this system to switch over to the appropriate ignition curve when the driver activates the fuel-selector on the dash.)

Why should a dual-fuel installation necessarily be any less efficient than a straight-fuel application? Aside from not being able to make use of propane-efficient modifications, the main problem is the carburetion. Essentially, in a dual-fuel installation the propane mixer is installed above the existing gasoline carburetor, and switching devices are incorporated so that only one fuel source and one carbureting device is used at any one time. What this does, in essence, is pile one carburetor on top of another, and as a result, both of the carburetors are less efficient.

Carburetors are "pressure-drop" devices. Conventional carburetors rely on the fact that the relative pressure in the engine manifold is less than the pressure at the point where ambient air enters the carburetor; and this differential, or pressure drop, between the manifold and the intake pressure causes inducted air to flow through the carburetor and pickup fuel from the nozzles in the main venturis. If the inducted air is forced to flow through two sequential carburetion devices (venturis), as it must in a dual-fuel installation, the pressure differentials across each of the carburetor (venturis) is weakened. During gasoline operation, the gasoline carburetor has to draw air through a restriction—the propane mixer, which is normally mounted above the gas carb. And during propane operation, the mixture leaving the propane mixer has to pass through the venturis and around the choke and throttle apparatus of the gasoline carburetor before it enters the manifold. As a result, the gasoline carburetor and the propane mixer will operate satisfactorily, but each is less efficient.

The result for most dual-fuel installations is a net loss of 5%-10% in efficiency, meaning that both power and economy will suffer. In effect, dual-fuel is a trade-off installation. The operator trades peak efficiency on either fuel for the flexibility in fuel usage and extended range.

From a mechanical standpoint a dual-fuel conversion is only slightly more complicated than a straight conversion. The major differences are: (1) the entire gasoline system is retained and (2) fuel lock-off valves are added to both the propane and the gasoline supply systems. Basically, the gasoline carburetor stays in the stock location, and the propane mixer is mounted directly above the carburetor. The air cleaner, usually not the stock one, is mounted to the propane mixer.

Propane mixers for dual-fuel systems are simple air-valve devices that provide the proper air/fuel mixture in response to negative pressure in the intake manifold. A Bowden cable (a slide cable similar to the old fashioned hand-choke cable) operates a cam-ring on the mixer to raise the mixer and open a direct air feed to the carburetor. The cable terminates inside the vehicle, where the driver can push in the cable knob to operate on propane or pull it out to lock the air-valve for operation on gas.

In addition, a dual-fuel system also requires a three-position electrical switch, which is usually mounted below the dash panel (along with the control knob for the air-valve control). This switch activates the two electric lock-offs in the fuel supply systems, and it effectively controls which fuel is supplied to the engine and prevents the possibility of both fuels being delivered to the engine at the same time.

The lock-off is simply an electrically-activated solenoid valve—one is installed in the propane liquid line just ahead of the converter and the other is put in the gasoline fuel line just ahead of the stock carburetor—and when the solenoid is not energized, fuel flow through the system is effectively cut off. The dash switch only supplies energy to one solenoid at a time. Generally the switch has a center-off position, and allows propane flow in the UP position and gasoline flow in the DOWN position.

In the captions to our step-by-step photo sequence, you'll notice that the three-way switch calls for an "unfused ignition" 12-volt energy source. This means the source must be OFF when the ignition key is off, ON when the ignition switch is on, and it must also be ON during cranking. Finding this source may take a little research because virtually all of the accessory circuits on late-model cars are switched off during startup. (This prevents the accessories from unnecessarily draining away battery energy during engine startup.) It is, therefore, vitally important to *positively ensure* that the energy source for the fuel switch is, in fact, energized during startup. If it is not hot at startup, both of the shut-offs will close and fuel will not reach the engine no matter what the switch setting may be!

In actual driving, fuel switching is relatively simple. If, for instance, the engine is running on gasoline, the air-valve cable is out and the switch is down. To switch to propane, the operator flips the switch to the middle (OFF) position, which shuts off the gasoline lock-off. The engine will run for a short period at highway speeds (longer at idle) on the fuel remaining in the float bowl of the carburetor. And just as the engine starts to die, the driver flips the switch UP for propane and pushes in the air-valve cable. To switch from propane to gasoline performance, the driver flips the switch directly to the gasoline position. When the fuel pump has filled the float bowl in the carburetor and the engine starts to run on gasoline, he pulls out the air-valve cable to prevent the air valve in the mixer from restricting airflow to the carburetor.

Both operations are best performed at steady highway speeds, when it takes less time to empty or fill the float bowl. It is also possible to switch from one fuel to the other while the engine is at idle, but it's best not to switch during stop-and-go traffic, when a momentary bobble of the engine could come at a time when the driver needs to accelerate smoothly.

If the operator desires a slightly more sophisticated system, a different type of fuel switch may be used. Instead of three terminals, it has six. In addition to the connections from the two lock-offs and the 12-volt source, it also connects to the stock fuel gauge. The fuel-level sensors in the stock gasoline tank and the propane tank are connected to the switch such that the fuel gauge reads from whichever fuel source the engine is using. Of course, the stock fuel gauge will still read gasoline levels even without the six-terminal switch and the propane tank has a level gauge (or more correctly, a percentage-of-fullness gauge) installed near the filler valve. The observant driver will always be able to tell how much of each fuel he has, but the six-terminal switch provides additional convenience.

In addition to the cautions we've already given about handling propane components, there's an extra-special caution to take on the gasoline portion of a dual-fuel conversion. Absolutely do *not* use a rubber fuel line or even a short length of rubber hose between the stock mechanical fuel pump and the gasoline lock-off. Use steel line with threaded fittings on each end. This sounds like a simple thing, but using even a short piece of rubber hose to make the conversion simpler *will almost guarantee a potentially hazardous engine fire.* When operating on LPG the gasoline lock-off shuts off flow to the stock carburetor, but the stock mechanical fuel pump is still pumping hard against the lock-off, and if conventional rubber hose is used anywhere between the fuel pump and the lock-off it may burst and spray gasoline all over the engine. All stock mechanical fuel pumps have an internal relief to protect the pump diaphragm when line pressure builds up, but this won't prevent a non-metallic line from bursting.

The pressure relief in the mechanical fuel pump can also pose another potential problem. It rarely happens, but if the relief failed for any reason while operating on propane, the diaphragm in the mechanical pump could rupture, allowing raw gasoline to pass into the crankcase of the engine. Unfortunately, the operator wouldn't know anything was wrong until the engine seized up from lack of lubrication. To eliminate this possibility many veteran propane installers highly recommend blocking off the stock pump and substituting an aftermarket electric fuel pump on any dual-fuel conversion. The electric pump can be hooked to a dash-mounted toggle switch or connected to the dual-fuel switch so that when the gasoline fuel lock-off is off, the pump is deactivated, relieving the supply line of excessive pressure.

The emissions equipment is potentially the biggest drawback to a dual-fuel installation. To be strictly legal, *all of the emissions equipment must be retained,* since the car can still run on gasoline. However, these devices and the extra internal and intake-tract heat needed for gasoline performance will hurt performance in the propane mode. Once again, you'll have to let your conscience and your local smog laws be your guide as to how you handle these modifications, because a professional conversion center is not allowed to remove or tamper with emissions equipment on a dual-fuel vehicle.

If you do decide, as many have in the past, to go dual-fuel with the intention of running almost exclusively on propane (with the gasoline system as a back-up), there's an additional caution to remember. When a gasoline carburetor is used only infrequently, the gaskets have a tendency to dry out and warp. When you finally do switch to "blacky carbon" fuel, the dried-out gaskets may allow fuel to leak from the carburetor and create a potential fire hazard. *For this reason most conversion centers and installers recommend that dual-fuel engines be run on gasoline at least once a week, even if only for a short period to keep the carburetor gaskets tight.*

Another reason for doing this on a regular basis is to keep the gasoline system in good working order.

In a typical dual-fuel conversion the LPG mixer is mounted atop the stock gasoline carburetor. This can sometimes create clearance problems, unless an offset adapter is used or the installation is in a truck that has a lot of hood clearance.

When gasoline is stored for long periods it will evaporate and produce a gummy "varnish" in the tank and supply lines, and this residue will inevitably cause carburetion problems. In addition, if the gasoline is used infrequently and the tank is kept only partially filled, there is a greater tendency for water vapor in the air to be absorbed by the gasoline. To avoid this problem it is a good idea to regularly use some sort of "dry gas" product in the tank. This "dry gas" chemical (Christy's Dry-Gas is one popular brand) will keep the water in suspension in the gasoline, and prevent it from collecting at the bottom of the tank, where it may be picked by the pump and passed into the carburetor and cause problems.

And as a final note we offer this last bit of advice for dual-fuel users. Though dual-fuel engines can be started up on either gasoline or propane, long-time users often recommend that the engine be started on propane whenever possible and warmed up to normal operating temperatures before it is switched over to gasoline operation. This takes maximum advantage of the gaseous nature of propane and it substantially reduces the oil dilution and cylinderwall wear that inevitably occurs when gasoline is inducted by a cold engine.

We're indebted for this photo sequence to the experts at the Petrolane Clean-Air Center in Santa Fe Springs, California. In addition to being one of the largest propane distributors in the U.S., Petrolane has an expanding network of highly efficient conversion centers that do all types of fleet and private-party propane conversions.

On this Chevrolet El Camino the tank will be mounted at the front of the pickup bed. Stiffening plates are used underneath because the bolt holes are located in relatively thin sections of the sheetmetal bed floor.

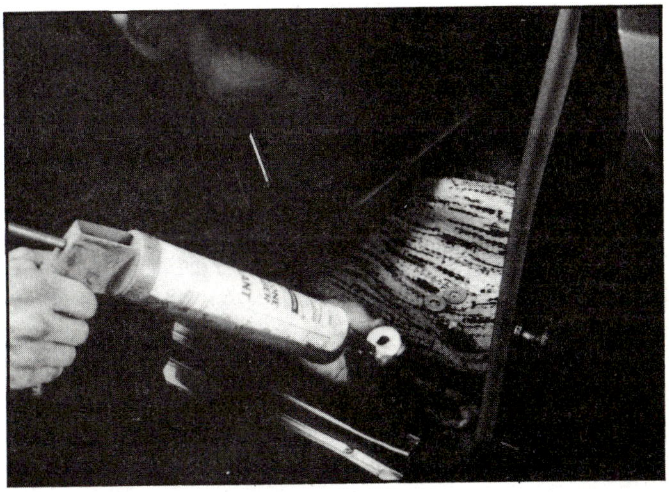

The mounting bolts go through the luggage space behind the El Camino seat. To prevent water leakage around the holes, silicone sealant is applied to the large fender washers used with the mounting bolts.

A three-way bulkhead fitting is installed in the truck bed near the tank. At right is the hydrostatic-relief valve, which prevents the line from failing in case of a sudden rise in line pressure.

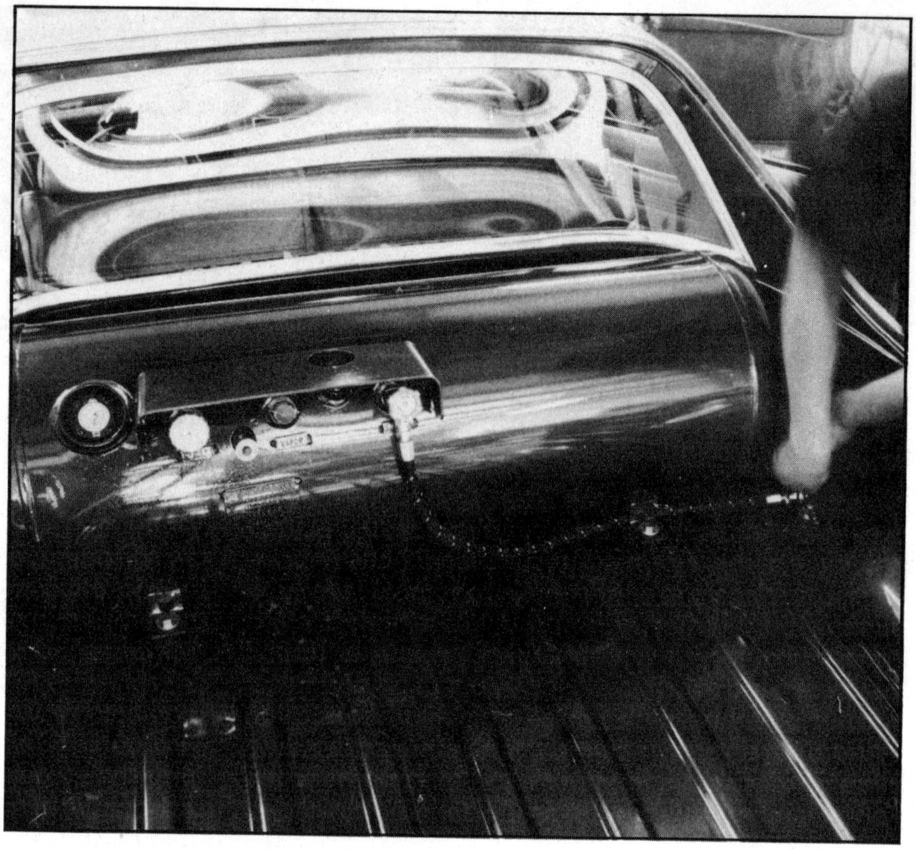

A high-pressure liquid line is assembled to connect the main liquid valve of the tank to the bulkhead fitting. A separate line runs from the third connection on the bulkhead fitting (beneath the bed) forward to the engine compartment. To reduce heat buildup from the sun, the tank should be painted white; but this one was custom-ordered the same color as the vehicle.

The line going forward should be securely attached with clips to the frame and routed away from exhaust heat or possible sources of chafing.

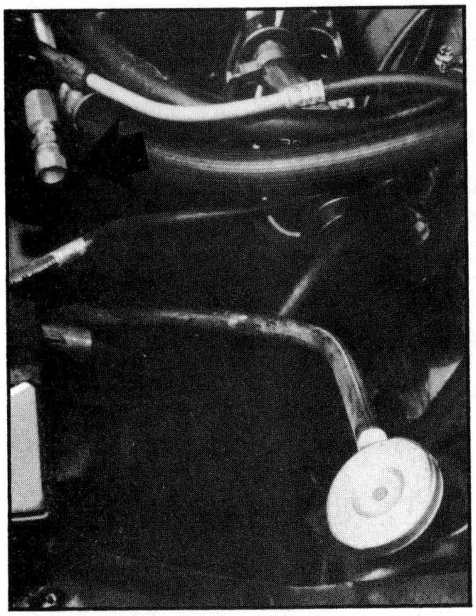

The liquid-fuel line is made just long enough (arrow) to terminate aft of the El Camino battery tray, where the vaporizer will be mounted.

The stock Rochester four-barrel gasoline carburetor remains untouched; only the stock air cleaner is removed.

The stock Chevy 350 is awash with smog equipment and accessory plumbing, but the conversion remains relatively uncomplicated.

The gasoline fuel line connecting the mechanical fuel pump to the carburetor is interrupted just forward of the carb and a lock-off device is plumbed into the gas line. The gasoline lines must be metal, not rubber. At lower left, note how the anti-backfire valve was relocated to the rear. Later, you'll see why.

The Impco 300A is a propane mixer designed especially for dual-fuel installations. It is simple and relatively compact.

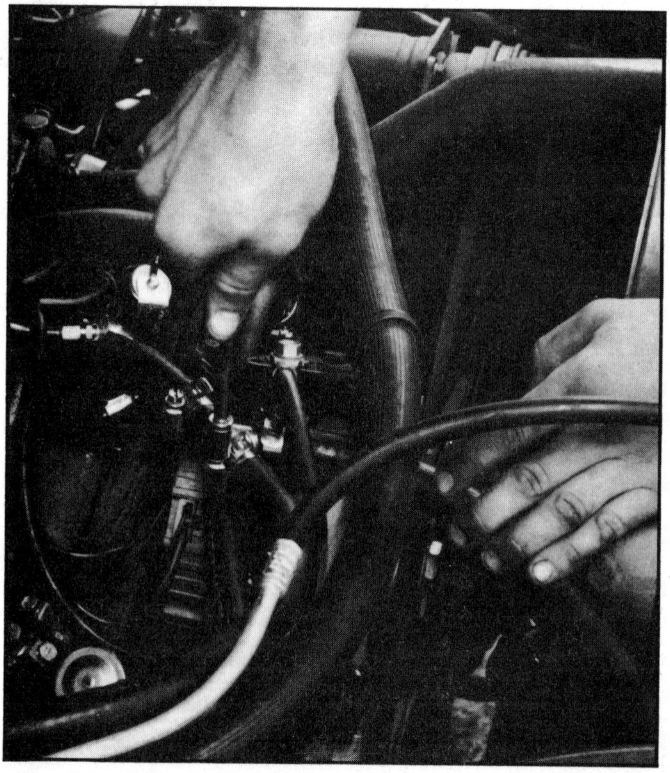

A pair of bronze Y-connectors from Gann Products are used here to interrupt the heater hoses and provide hot coolant to the Impco vaporizer.

Viewed from the bottom, the simplicity of the dome-shaped mixer is apparent. A cable-operated cam-ring either locks or unlocks the mixer diaphragm, depending on whether the engine is operated on gasoline or propane.

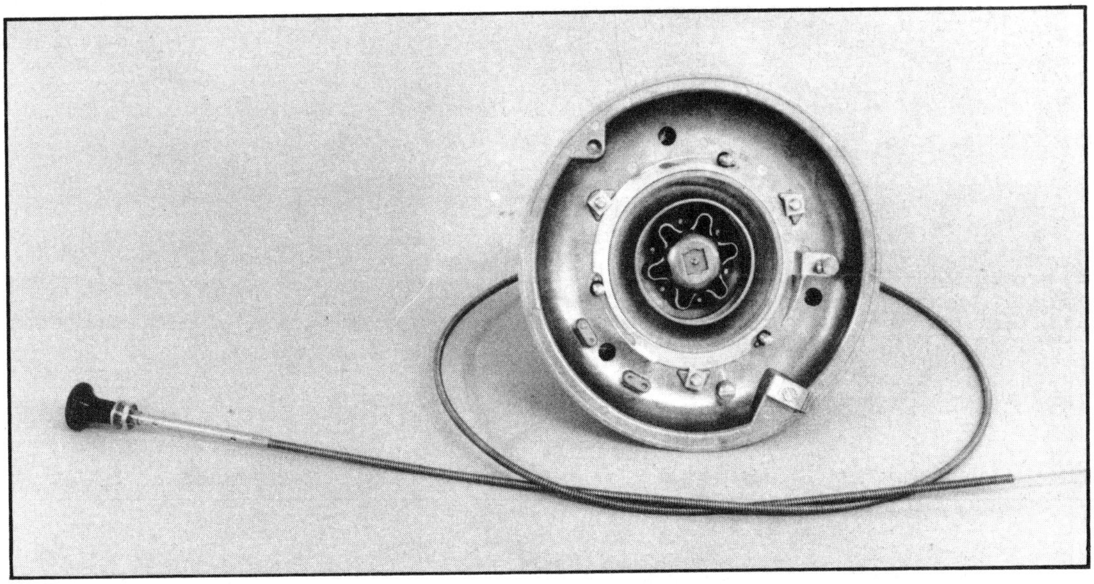

The arrow indicates where a plug normally covers the idle-mixture screw. To the left of this threaded hole is the power-mixture adjustment. The white sticker specifies which model vaporizer is recommended for this mixer.

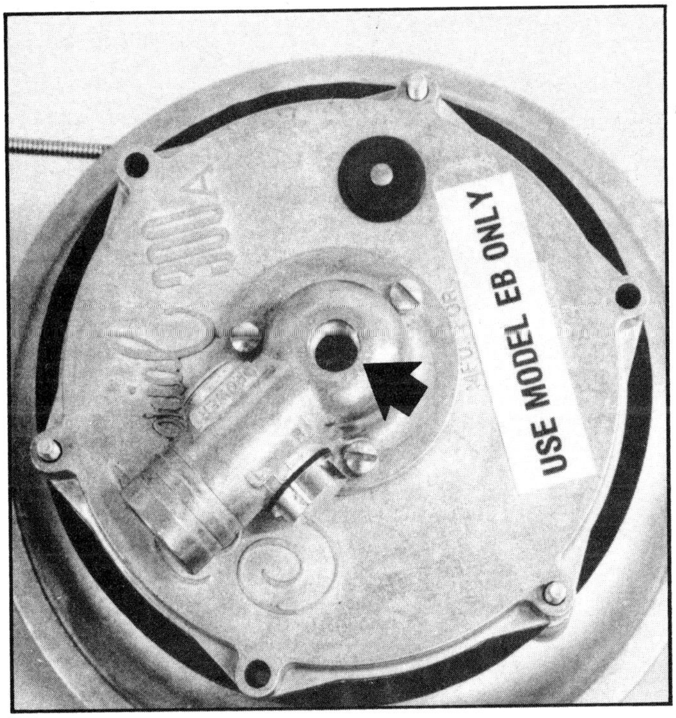

After the Bowden cable is routed from the dashboard to the engine compartment, the outer cable is secured by a clamp and the wire tightened with the cam-ring turned counter-clockwise as far as it will go.

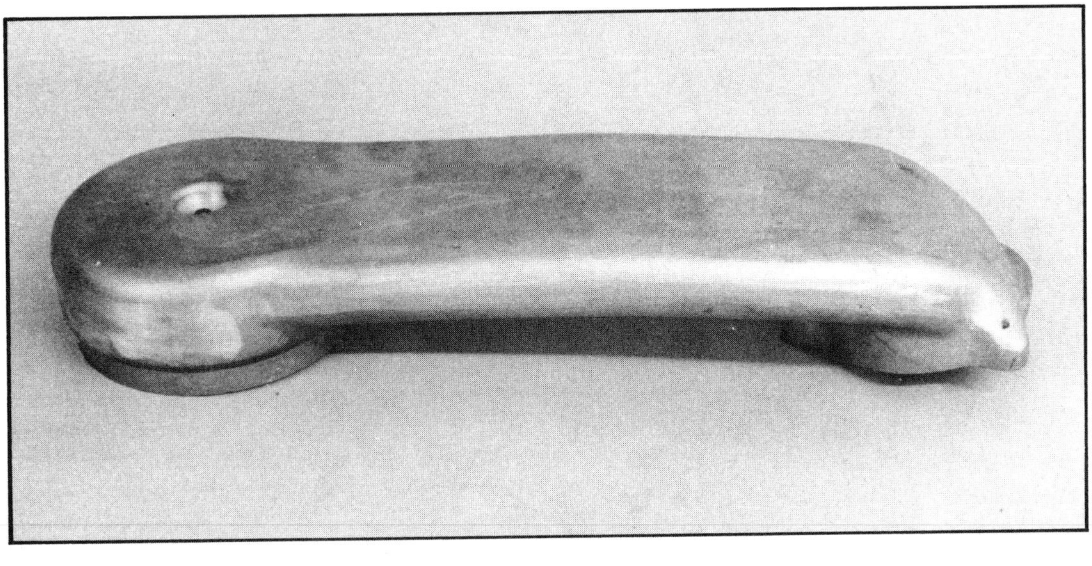

A long, offset aluminum carburetor adapter was necessary on this El Camino. Due to limited hood clearance, the 300A mixer could not be mounted directly above the Rochester carburetor.

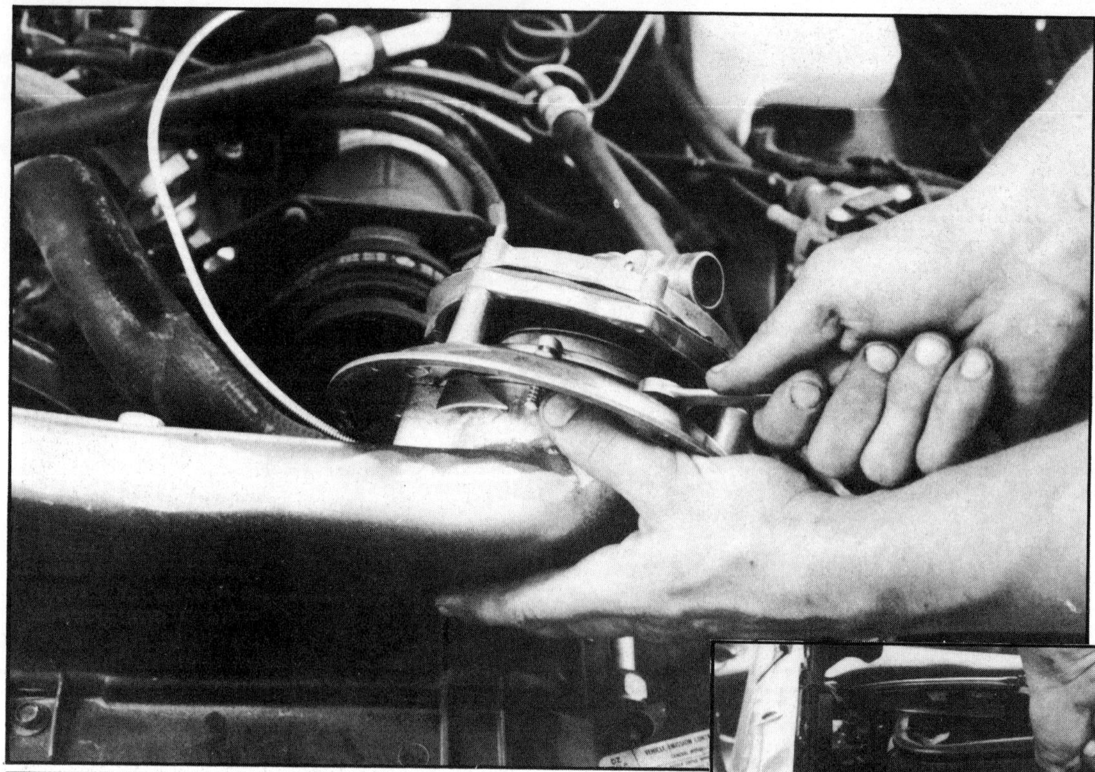

The propane mixer bolts to one end of the aluminum offset adapter. A standard air-cleaner element is slipped on, and then an Impco air-cleaner ring is bolted in place. In this compact package the mixer is contained almost totally within the air filter. Here, the propane vapor hose is attached; the other hose is for the stock PCV system.

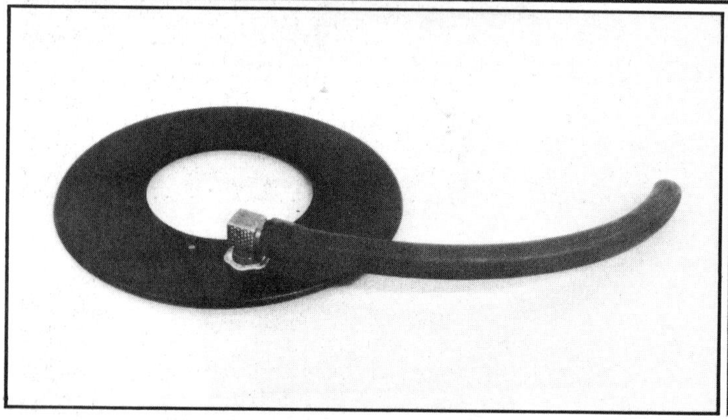

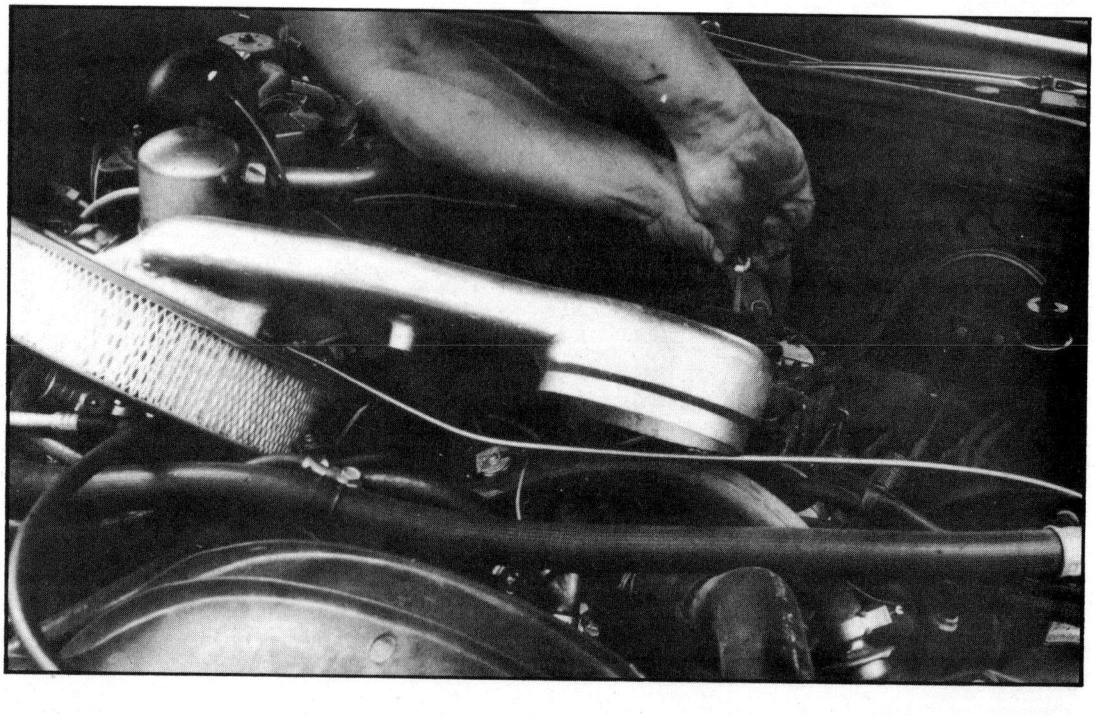

A longer carburetor center stud is screwed into the Rochester gasoline carburetor so the aluminum offset adapter can be fastened in place.

An assistant works the Bowden cable back and forth a few times while the installer watches the cam ring inside the mixer to make certain it operates properly. Sometimes the aluminum adapter needs to be relieved (arrows) with a file or grinder to clear the diaphragm screws and the cam ring.

A steel plate simplifies mounting the Impco Model-E vaporizer (also called a converter), shown here with all the fittings in place. At left is the Beam Products 12-volt electric lock-off for the liquid propane. It shuts off the liquid whenever the ignition key is in the off position, just like a standard electric filter/fuel-lock; but it is also wired to shut off the liquid supply when the dash-mounted switch in a dual-fuel installation is turned to the gasoline position.

The steel plate is bolted securely with locknuts to the inner fenderpanel of the El Camino, and the heater hoses are hooked to the vaporizer.

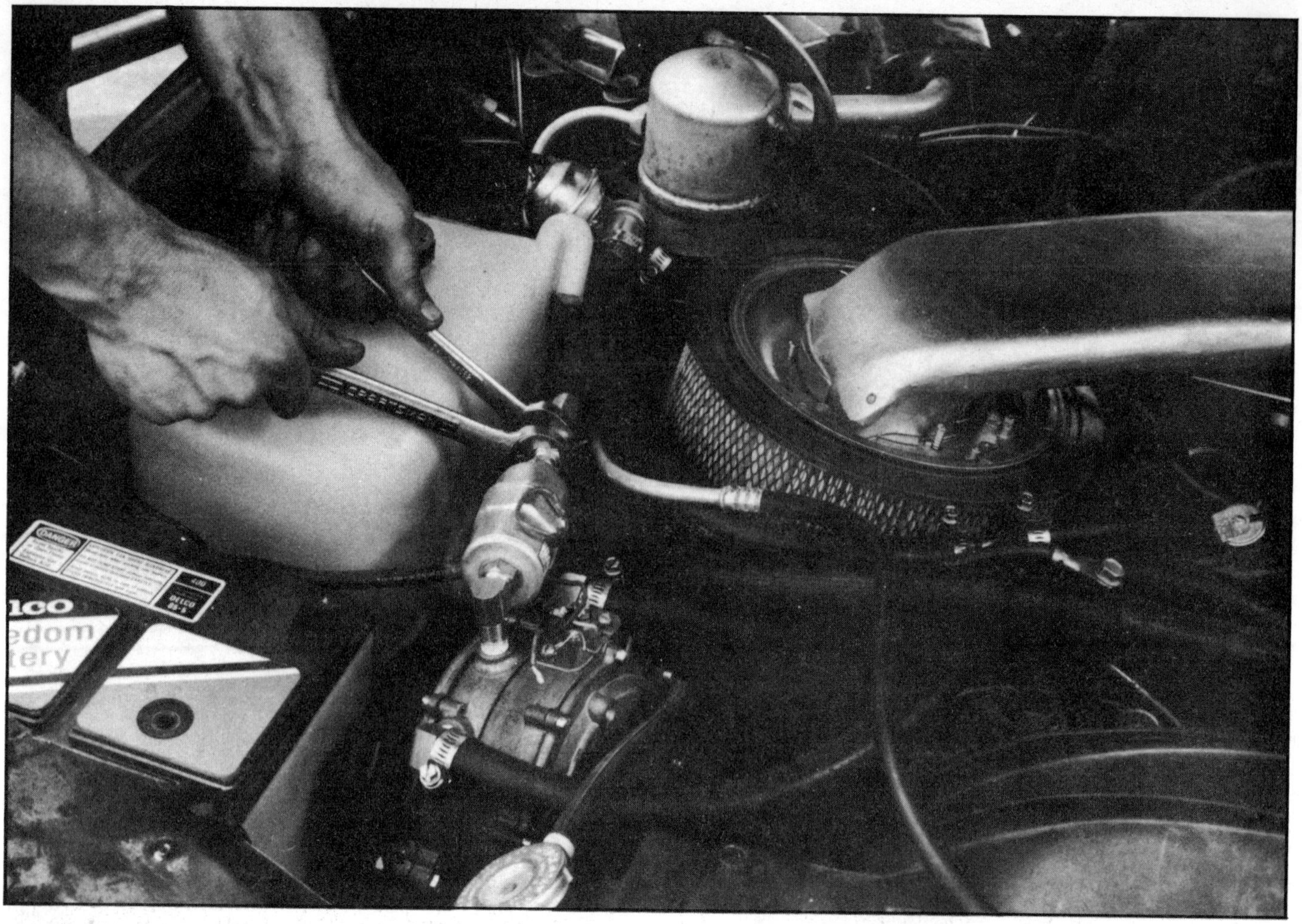

The high-pressure liquid line from the tank is now hooked to the filter/fuel-lock. The anti-backfire valve, which was relocated earlier, is moved to provide clearance for the new mixer/air-cleaner position.

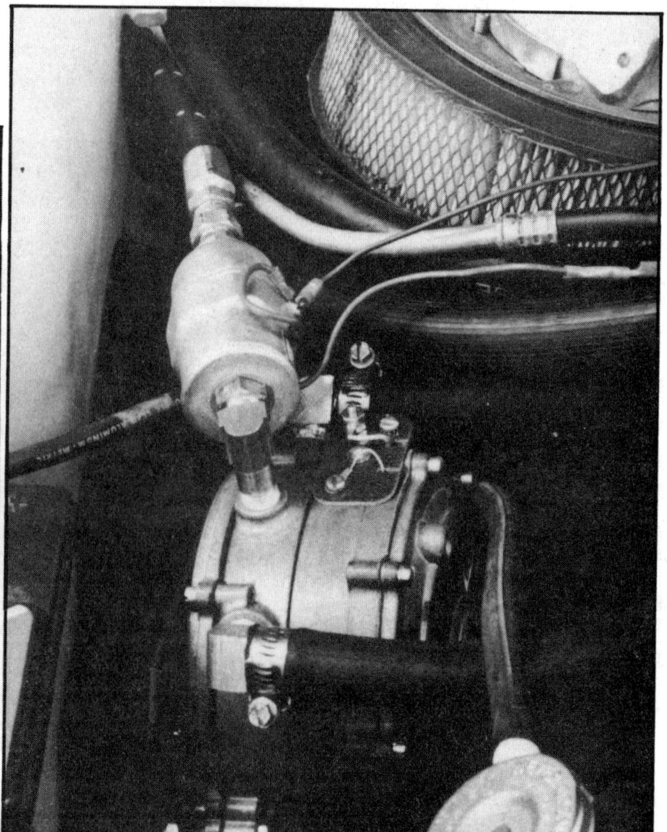

Both of the electric fuel lock-offs are now wired to the switch panel under the dashboard, and a ground wire joins the two solenoids.

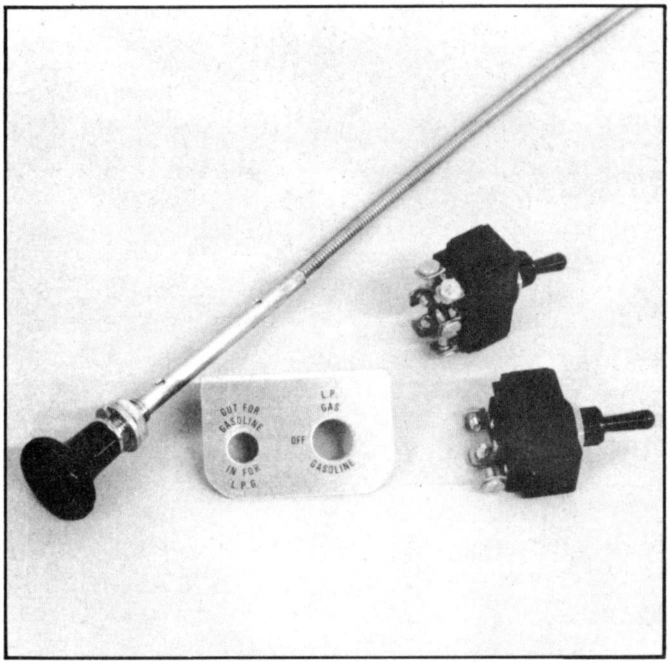

The Bowden cable mounts inside the vehicle in conjunction with a three-way electrical switch connected to the solenoids—for selecting either propane or gasoline operating modes. Using the six-connector switch, rather than the three-connector, allows the stock fuel gauge to read either gasoline or propane, according to the mode of operation.

Any location under or in the dash is suitable, as long as it is convenient for the driver to reach the controls.

After the engine is running again, the Petrolane technician checks all liquid connections with a Manchester leak detector. The El Camino is now ready for final checkout on the chassis dyno.

CHAPTER 5 TURBOCHARGING

PROPANE AND TURBOCHARGING

Turbocharging used to be discussed more often in racing circles than by drivers interested in the economy and utility of a vehicle. This is happily not the case today; there is now a steadily increasing interest not only among general motorists but even in the utility-minded hallways of Detroit's automakers. Forced by governmental pressures and foreign manufacturers to increase fuel economy, Detroit has recently acknowledged what racing enthusiasts and aircraft designers have known for many years: a turbocharged engine can deliver power that belies its size. A small, efficient engine with a good turbocharging system can produce the peak horsepower of a larger engine, without a comparable increase in overall fuel consumption!

Without question, the adaptation of turbo principles to the rigors of daily driving has been a slow process, pockmarked with false starts and some well-known failures; but the widespread acceptance that turbocharging enjoys among automotive engineers speaks volumes about modern turbo systems. In recent years, the number of turbocharged cars has proliferated dramatically. There are currently about 40 domestic and imported models offered in turbocharged form, compared to only 9 in 1980. And it now seems that the word "turbo" has become one of the top catch-phrases of the Eighties. It is used by zealous marketeers to give a modern, high-tech connotation to everything from clothes to computer games and aftershave lotion!

To understand how this magic works, let's describe what a turbocharger is and how it functions. Not unlike the piston-engine itself, the turbocharger is an *air pump*. Basically, it is composed of two fan-like wheels, called turbines, which are inside housings that fit closely to the blades on the turbines. Each turbine wheel is connected to the other with a common central shaft. One of these wheel-housing arrange-

Few know more about the combined advantages of propane and turbocharging than Ak Miller. Conversions at Ak Miller Enterprises run the gamut from big trucks and motorhomes to high-performance street machines. This 1982 Capri with a Ford 351 W V-8 engine and twin turbos exhibits excellent all-around driveability, immense horsepower, and 15mpg economy on propane.

Ak Miller is a long-time veteran of racing at Pike's Peak, Colorado. Several of his race vehicles have run the course on propane. This example is a sprint car powered by a 351-C Ford with twin AiResearch turbos and LP carbs. The altitude compensating characteristics of this setup makes turbo/propane an ideal combination for the "Race To the Clouds," yet it's equally suitable for trucks and motorhomes.

The typical turbo-propane truck installation, while less glamorous than Miller's racer, is nonetheless efficient. Changing to a lower-numerical gear ratio in this 428 Ford pickup improved mileage since the turbo allowed the engine to work better at lower engine speeds.

Installed on a Pinto, the above Ak Miller kit turned in some amazing performances at the Bonneville Speed Trials.

ments is called the exhaust turbine. This wheel is driven by the flow of hot exhaust gasses escaping from the engine. These gasses flow over the wheel, causing it to spin at a very high rate. When the gasses exit the housing, they pass through the exhaust pipe and muffler. Since the two wheels are connected by the common shaft, the other wheel, called the compressor, is driven at the same high speed. The primary function of the compressor section is to force air or air-fuel (as the case may be) into the intake tract of the engine.

The difference between a normally aspirated (non-supercharged) engine and a turbocharged example is somewhat like the difference between a mountain climber breathing thin air with sea-level lungs and another climber breathing with an oxygen tank and mask. In a simplified sense, a turbocharged engine is *force-fed* an air-fuel mixture, instead of having to "suck" it in.

From an engineering standpoint, volumetric efficiency is the key to engine power: packing the maximum amount of air-fuel into the cylinders and doing the most with what is packed in. The turbocharger radically improves the volumetric efficiency in a way that traditional performance tricks (such as larger valves, head porting, radical cams, and intricate induction systems) can't.

The true beauty of a well-balanced turbocharger system is that it creates a *variable-horsepower* engine that can be extremely docile until the driver asks for something more. A turbo system responds on engine *demand* and *load*. Therefore, the engine can be smooth and economical until a heavy load is imposed on the vehicle (such as pulling a trailer or camper,

A cutaway of an AiResearch TO4 turbocharger illustrates the relationship of the two turbine wheels, the common shaft, and the central bearing. On the left is the turbine or exhaust portion; at the right is the air compressor. Obviously, to run at speeds of up to 100,000 rpm, the components have to be built of special materials and to close tolerances.

A turbocharged propane engine loves to work hard, and the advantages of propane fuel allow it to last much longer (under load) than conventional gasoline engines. This Oldsmobile V-8, used as a stationary power source for an industrial pump, was designed to operate on LPG; but it was retrofitted with a turbo system to produce the same horsepower at a lower engine speed, thereby increasing longevity.

climbing up a steep hill, or needing rapid acceleration). When this happens, the driver opens the throttle, the flow of exhaust gasses increases, the turbine wheels in the turbo spin faster, and the compressor really packs the intake of the engine. However, when the load is light, exhaust flow is decreased, the turbines spin more slowly, the intake is not "supercharged," and the fuel consumption is not adversely affected.

Fuel economy is a matter of interest for most drivers, and propane users are just as touchy about the subject as everyone else. And as hard as it is to believe, turbocharging can prove of substantial benefit *when used properly.* In normal driving, there may be little difference in engine efficiency—and if the driver can't keep his foot off the floor, there may even be a loss of efficiency. But the key to mileage in any installation is *the rpm range of steady-state driving.*

Let's take a typical example. Joe Weekender has a late-model Ford pickup with a camper; and on weekends, he likes to take his family to the wide-open spaces. His truck has a stock 390cid V8 engine and a 4.11 rear-axle ratio. At 55mph the economy is about 9 or 10 miles per gallon. He's not very happy with the mileage, and he's heard something about turbochargers; so he spends some money for a turbo kit and installs it on the 390. The next weekend, he's ecstatic about the improvement in performance, especially on those long grades; but he doesn't see much improvement in mileage. So he gets some good advice from a turbo expert, and he installs an overdrive or changes the rear-axle gearing to something like 3.23 or 3.00:1. The next weekend, he gets the pocket calculator out after a trip to the desert and finds the mileage is up to 12-14mpg!

Without the turbocharger, the engine was heavily loaded on hills, especially during towing, and a low rear-end ratio (the numerically high 4.11's and the higher engine rpms they produced), was required to get the job done without having to downshift to second gear. After the turbo was installed, the new 3.23 gears provided better level-road economy, *and the turbo provided enough load-demand power to tackle those long grades.* Without the extra torque from the turbo available for hill climbs and other "performance" applications, the truck would not be practical with the "economy" 3.23 gearing. Under boost, the turbo provides a 7 to 10% increase in torque for each psi of boost pressure.

So a turbocharger can help an engine *work* more efficiently, but only when the engine is properly geared to take advantage of the turbo capabilities.

Although uncommon, propane can also be used successfully with mechanically-driven superchargers. This unique supercharged and propane-fueled smallblock Chevy was built by Don Bass.

The blown-propane Chevy was used in a ski-racing boat that was the scourge of the waters until it was outlawed. It could run with the best of the nitro-fueled big-block engines.

This is the guiding philosophy behind most big industrial engines. Generally, turbocharged and running on natural gas, these big V-12s and V-16s are designed to provide the necessary horsepower for their job at only 1000 to 2000rpm. They work efficiently, and the lower piston and bearing speeds afford these expensive behemoths tremendous reliability and life.

Turbocharging is not an inexpensive option and certainly not "necessary" in most installations; but it works so well with propane, it seems a shame not to take advantage of the combination in cases where there is a workload to handle or simply when you want more performance without affecting engine smoothness, reliability, or fuel economy.

If this all sounds like the ideal way to build an engine, then read on because *turbocharging and gaseous fuels (such as propane and natural gas) are ideal for each other.* Of course, there are some drawbacks to turbocharging. The main drawback is the increased tendency for the engine to detonate, However, you just happen to have 105-octane fuel in the propane tank! For maximum efficiency, the increased pressure in the combustion chamber of a turbocharged engine (especially during extreme-boost conditions) requires a fuel with excellent anti-knock properties (high octane) to reduce detonation difficulties. It's for this reason (and others) that propane is ideal for a turbocharged engine.

Current pump gasolines are not really suited for turbocharged engines because of the drastic decline in octane ratings in recent years. To put is simply, the heavier the fuel, the lower the octane rating, unless a suitable chemical (some sort of anti-knock agent) is added to boost the octane. In bygone times, the availability of premium gasolines with tetra-ethyl lead additives (an excellent anti-knock agent) made it possible for designers to develop efficient, high-compression engines; but environmental concerns have forced oil companies to phase out leaded gasolines. And despite tremendous research, petro-chemists have not discovered a more suitable anti-knock agent for gasoline than tetra-ethyl lead. The only alternative has been to increase the octane rating by further refinement of the gasoline into lighter forms, which is very expensive. As a result, the simplest solution (compromise) has been to alter the design of automotive engines to accept fuels with lower octane ratings; therefore, current production cars have very low compression ratios (usually about 8:1) and very little spark advance.

It's interesting to note here that one of the methods used by some oil companies to improve the octane ratings of unleaded gasoline is to "spike" the fuel with butane (91 octane). If the refinery happens to produce a batch of gasoline that tests below the government requirements (at least 87 octane), low-pressure liquid butane can be added to raise the octane rating. The government only tests the product as it leaves the refinery; so three or four weeks later, after the gasoline has been in the storage tank at the local gas station for awhile (and most of the butane has evaporated), the octane may be lower than the minimum requirement.

There are other reasons why propane is a very suitable fuel for a turbocharged engine, not the least of which is the very good *homogenization* of the inducted propane-air mixture. As we noted with our earlier discussion of propane benefits, it is difficult for most engines to fully vaporize gasoline. Under less-than-ideal conditions, some of the gasoline enters the intake manifold in the form of small droplets that are simply pushed toward the combustion chamber by the incoming airstream. Along the path through the intake manifold and cylinder head ports, these droplets can readily "fall out" of the airstream. Inevitably, these droplets collect into little puddles; and as the puddles swash around in the manifold runners, they eventually reach one or the other of the combustion chambers. This means that some cylinders will randomly receive a gulp of very rich mixture, and fuel distribution (the relative amount of fuel delivered to each cylinder) will be very uneven.

This inherent difficulty with uneven fuel distribution can be a very severe problem in a turbocharged engine, because the carburetor may not be mounted on the intake manifold (as we'll see later, it may be

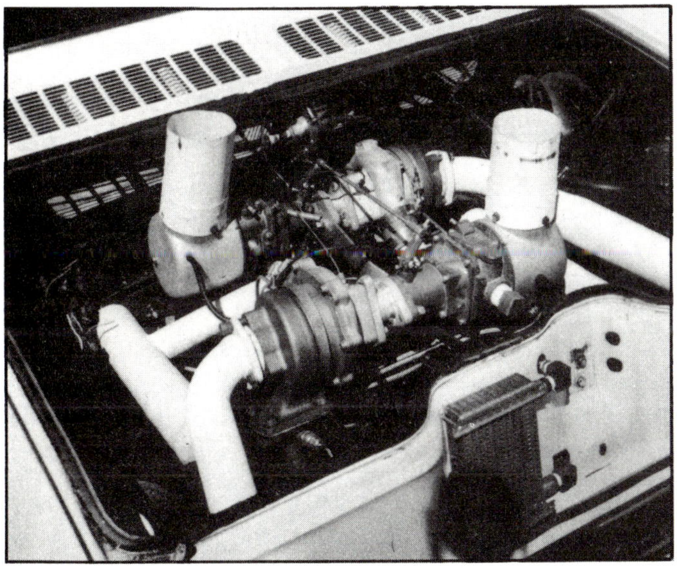

Propane has been used in a number of different types of race cars; but one of the most interesting was Tom Keosababian's 1965 Corvair, which set a new record of 173.09mph at Bonneville. This is a considerable achievement; but it's truly impressive when you consider that Tom's Corvair is an unmodified, street-driven chassis using the original air-cooled engine! Puffed up with two Rajay turbos and fitted with two Impco Model-E vaporizers and Model CA425 mixers, the GM flat-six developed 450 horsepower on the dyno.

much further away), leading to even greater "dropout" difficulties. But, obviously, running on propane vapors eliminates fuel puddling and distribution problems. And when distribution problems are completely eliminated, certain types of turbocharger installations are much more practical and efficient!

In many ways, turbocharging and propane are naturally complementary: each increases the benefits of the other. Earlier we noted that there was about 34% less BTU content in propane (when compared to conventional pump gasoline), and we should therefore expect a 34% drop in efficiency and power when converting to propane However, we know from experience that a real-life engine only loses about 5% to 15% efficiency when it is converted to propane (all else being equal; the difference is largely due to the highly homogenized intake mixture produced by the gaseous fuel system.

In a naturally aspirated engine, this 5% to 15% loss

The author's latest street-rod project is a 1932 Ford coupe powered by a 302 Ford V-8. It's not very difficult to make a lot of horsepower with standard performance techniques, but there's a special satisfaction in making a fast, efficient vehicle that isn't a heavy polluter. In this setup, stock exhaust manifolds were swapped side-for-side and turned upside down. They were then joined with a crossover that mounted the turbo high and to the rear of the engine, where it pressurizes the propane mixer.

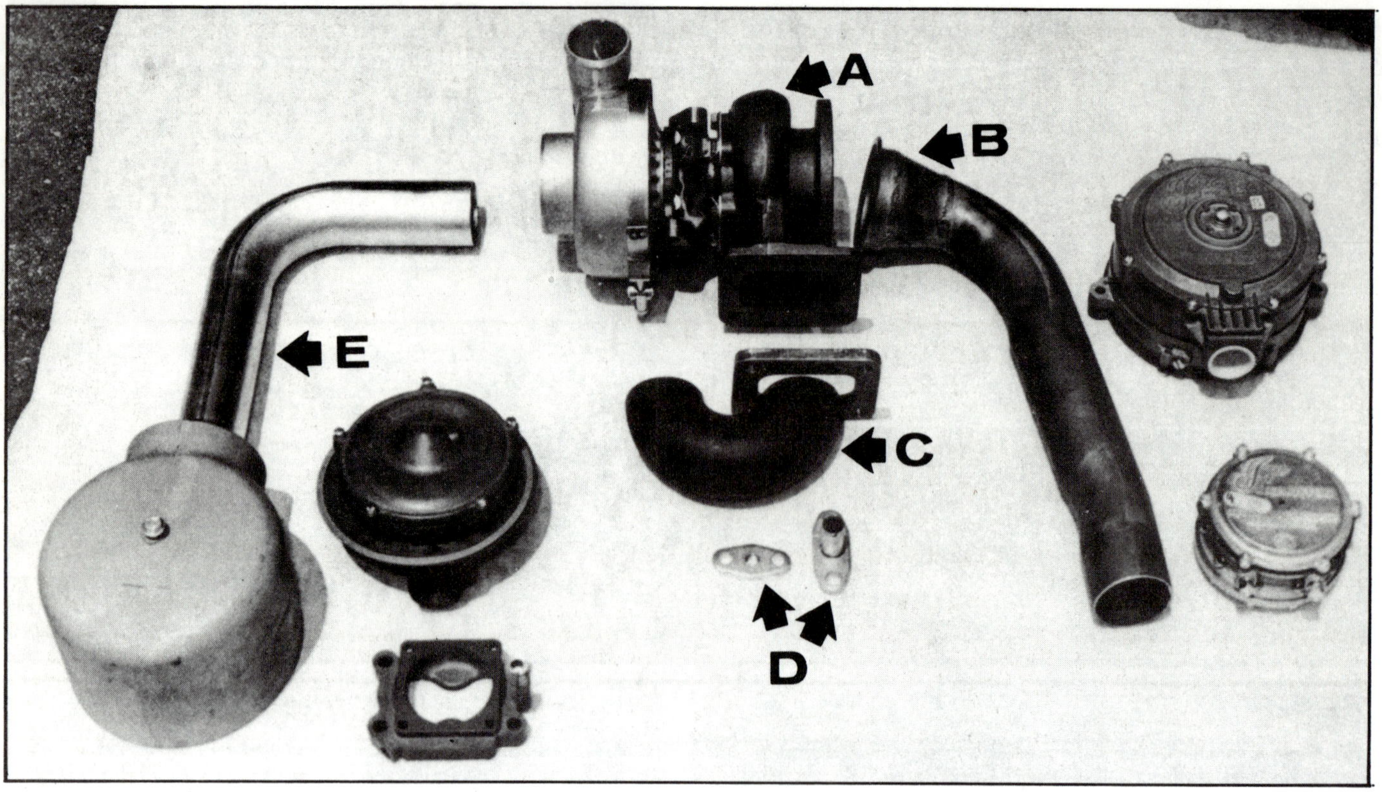

In addition to normal propane carburetion equipment, a turbo-propane conversion will require: (A) a turbocharger that has been selected according to the engine and usage, (B) a connection from the turbo exhaust outlet to the stock muffler system, (C) a connection from the stock exhaust manifold to the turbine, (D) supply and drainback fittings for turbo lubrication, and (E) a connection from the compressor outlet to the carburetor. This example is a typical blow-through system.

of power is inevitable; but with a turbocharger, *the lost power is recovered, along with a gain of about 40% to 50% more power!* Do we need to give you any more reasons why these two pair up so well? We've explained elsewhere how an optimized engine can recover some of the lost power (due to the lower BTU-content of propane). Well, a turbocharger is probably the easiest way to optimize the engine and take full advantage of the relatively high octane rating of propane.

A non-turbocharged propane engine should have more than 8:1 compression (8:1 is typical of most late-model Detroit production engines); but raising the compression will normally require installing expensive, high-compression pistons. But 8:1 is about the perfect ratio for a turbocharged propane engine! So, in many cases, the builder won't have to modify the short block or the cylinder heads to optimize the engine. In addition, the hardened valves and valve seats used in most late-model, low-compression cylinder heads (for unleaded fuels) are very suitable for propane.

A great deal has been written about turbocharger installations, much more than can be recapped here in just a few pages. However, we can give you the basics and illustrate some of the more common examples of turbocharged propane engines. (*Super-*

It is a lot easier to locate a source that can provide a complete turbocharger kit for a specific vehicle. The kit manufacturer has already devoted the development time to find the correct turbo size for the specific application. This is an Ak Miller kit for a Ford two-liter four; the turbo mounts on the exhaust manifold (far side) and blows through the propane carburetor.

Power, from S-A Design Publishing Co., provides more detailed information about supercharging and turbocharging, see page 128.) Built to withstand extremely high temperatures and the effects of very high shaft speeds (turbocharger impellers spin as high as 100,000 rpm), turbochargers are simple in design. Nonetheless, they must be manufactured

Inline engines are particularly well-suited to turbocharging, especially when both intake and exhaust manifolds are on the same side of the engine (the plumbing is generally less extensive). This International Harvester six is a blow-through installation, with the compressor blowing intake air to a mixer mounted in place of the stock carburetor. Generally, no modifications are necessary to use an LPG mixer in a blow-through setup.

from specialized materials and finished to very precise machine tolerances. As you might expect, this makes them rather expensive. A basic turbocharger suitable for automotive engines will cost about $350 to $450, depending on the brand name and the supplier. In addition, a lot of specialized plumbing and fabrication will be required before the system is complete and functional. So make no mistake about this, unless the installer is unusually ingenious and an experienced fabricator, a top-quality turbo installation can be formidable and expensive!

Installing an existing prefabricated kit is the only practical way to turbocharge most engines. Buying a used turbo at a swap meet and bolting it onto an engine will inevitably produce one of two results: 1) the engine won't run any better than it did stock, or 2) you may turn a freshly-rebuilt motor into next week's trash. When you buy an existing kit from a reputable source,

This custom turbo-propane installation in the author's 1932 Ford roadster street rod uses two small Impco mixers—for better distribution—on a homemade log manifold. The 300cid Ford six with C-4 automatic ran 12.90's at the drags in street trim. The engine provides plenty of power, excellent reliability, smooth operation, and much lower emissions than a modified gasoline engine.

they have already done the experimenting and engineering to develop the best turbocharger profile for your specific engine size and application. Such factors as engine displacement, carburetion airflow, underhood space restrictions, exhaust system backpressure, gearing, and intended usage should all be taken into account when selecting a turbocharger.

Good kits cost about $1000 to $2000 and generally include everything you need to make the conversion: special exhaust manifolds, carburetor adapters, hoses, linkages, and often a special air cleaner. Notice we said exhaust *manifolds*; this refers to conventional cast-iron manifolds, not steel tubing headers. Since thick cast-iron retains more heat than the lightweight steel tubing using in performance headers and since heat and pressure are what make the

When the engine must pull heavy loads or exceptional performance is desired, two turbochargers can be used. Note the turbo on each exhaust bank of this big Ford V-8, with each blowing into an Impco CA425 mixer.

turbocharger perform, cast-iron manifolds are used in turbo applications. (Usually just one manifold is supplied in kits, with a mounting flange for the turbine housing. The installer must provide a stock manifold for the other side of a V-8.)

On inline engines, the turbo usually mounts on a special manifold (normally provided with the kit), and new exhaust plumbing extends from the outlet side of the turbine housing down under the vehicle to hook-up with the exhaust system. On V-type engines with a single turbo (dual-turbo installations are expensive and not very common!), the exhaust from one (stock) manifold is plumbed under the engine to the other manifold—which carries the turbo—so that *all* of the exhaust gasses must pass through the turbo and turn the turbine wheel.

The outlet side of the turbo (exhaust) is large, to reduce backpressure; and it's important that the rest of the system be equally free-flowing. The most desirable kits will include a complete exhaust and muffler system that will complement the performance of the turbocharger. If the kit doesn't include an exhaust system, a suitable system will have to be fabricated. Most muffler shops can perform this type of work. The tubing should be at least 2.5 or 3.0 inches in diameter and a low-restriction muffler is essential. The original Corvair "turbo" muffler is very desirable, but it is no longer produced. However, several aftermarket manufacturers market namesakes that are supposedly as good as the original GM model. But beware of cheap replicas; they may not be as efficient or as durable. (Among the better turbo-type mufflers discovered in recent performance-magazine comparisons are: the American Exhaust Industries' "Sonic Turbo," the Maremont turbo replica, the Supreme "Super C Turbo," and the ubiquitous Midas turbo replica.)

Do not simply hook the turbo up to the existing ex-

For this dual-turbo installation Ak Miller used a separate vaporizer on each carburetor to allow sufficient airflow in high-demand situations. Hauling up a long grade, even loaded with an RV trailer, this rig can pass just about anything!

A specialty intake manifold could have been used with the two mixers; but since hood clearance was sufficient, an adapter was used to mount both mixers over the stock 4-barrel manifold.

In this Toyota installation Ak Miller connected the compressor outlet to a simple throttle body mounted in the same location as the stock carburetor. Connections must be made with large clamps and reinforced rubber hose.

haust system and muffler! The inherent restriction in most stock exhaust systems will limit maximum boost level and cause the turbo to develop boost more slowly. And one last bit of advice: if someone fabricates a custom system for you, instruct them to make as few bends as possible; and what bends have to be made should have the largest radius possible—no sharp kinks.

There are two basic ways to integrate a turbocharger into the induction system: either as a *blow-through* or as a *draw-through* configuration. Each has its own peculiar advantages and disadvantages; but once again, no matter which you choose, there is a distinct advantage to using a gaseous fuel, like propane.

The two designations refer to where the carburetor is mounted in relation to the compressor side of the turbo. In the blow-through mode, the turbocharger draws atmospheric air from the air filter and blows it into the carburetor. In the alternative design (the draw-through method) the carburetor is mounted in front of the compressor housing. The turbocharger draws both air and fuel through the carburetor, and the mixture is then blown directly into the intake manifold.

The draw-through method is very common with gasoline engines because gasoline carburetors are not designed to be pressurized. Conventional carbs react to atmospheric pressure, not boost pressure. Nonetheless, most gasoline carburetors can be modified, with considerable effort, to work adequately in a blow-through system. The throttle-shaft bores must be modified to prevent fuel-air leakage (several methods are available, but all require some fabrication and machine work). The fuel-bowl floats must be able to withstand the boost pressure (conventional hollow floats used in many carburetors are unsuitable, and must be replaced with solid, Nitrofil-type floats). The fuel-feed pressure must be augmented to overcome the boost pressure in the float bowl (an auxiliary electrical pump is usually suitable). And all of the conventional vacuum connections (i.e.,

Propane carburetion adapts to both blow-through and draw-through turbocharging systems more readily than gasoline carburetors. This Toyota mini-truck uses a draw-through with the mixer mounted directly to the compressor inlet. A gasoline carburetor would require an adapter for upright mounting and hot-water lines for proper vaporization.

This Toyota Land Cruiser off-road machine was pretty anemic before the turbo-propane conversion; now it acts like a big V-8, while retaining excellent overall economy. The turbo draws through two air cleaners and blows into an aluminum "hat" over the propane mixer. Adequate air filtration is always important in dusty off-road environments; even more so with a turbocharger.

Trucks are common turbo-propane candidates because they are frequently asked to pull loads, such as boats, campers, or RV trailers. Note that this Ford blow-through has a Mallory ignition coil and Ford silicone spark-plug wires and boots. A top notch ignition system is not optional; it is a virtual necessity on a turbocharged engine.

straight manifold vacuum, not ported vacuum lines) and vapor-recovery connections to the carburetor must have one-way check valves installed in them (or they will receive boost pressure instead of vacuum). However, despite these drawbacks, the major advantage of the blow-through system is that the carburetor remains in the stock location; therefore, throttle and transmission linkages do not have to be altered—and there is still a suitable, and essential heat supply under it to aid in fuel vaporization.

A simple variation of the blow-through system avoids some of the extensive carburetor modifications mentioned above. In this type of system a sealed pressure-box surrounds the entire carburetor, and boost from the compressor is directed into the box—and thus into the mouth of the carburetor. With this approach, the carburetor does not have to be modified to prevent the loss of boost pressure. But while internal modifications of the carburetor are eliminated, there is still the problem of routing lines and linkages into the box (while retaining a tight seal). Of course, you also have to take the pressure-box apart or remove

After a blow-through turbo system is installed, some trucks don't have adequate hood-to-engine room for a stock air cleaner (even when a low-profile intake manifold is used). In most cases, the compressor intake can be plumbed with flexible, reinforced hose to a remote air cleaner (left).

the carburetor from the box whenever you wish to work on it.

On the other hand, with a draw-through turbo system the gasoline carburetor is usually mounted on a separate manifold that attaches directly to the inlet side of the turbo compressor housing. As a result, all associated vacuum/fuel lines and linkages have to be relocated. A chamber must also be provided under the carburetor pad so that hot coolant can be channeled

One of the attractive engineering features of the OHG carburetor is the vacuum-triggered "economy valve." Comparable to the accelerator pump on a gasoline carb, this valve (arrow) opens when manifold vacuum drops below four inches and delivers an extra shot of fuel. This improves "driveability" and makes the mixer especially suited to turbocharged applications.

through the carb manifold to gain satisfactory fuel vaporization. In addition, a large-diameter tubing duct must be fabricated and high-strength hoses and clamps must be obtained to complete the connection from the output of the compressor to the intake manifold—which now sports a simple inlet adapter instead of a carburetor. Although there is sometimes a problem in finding room for the turbo and the carburetor (in the new location), the draw-through system usually doesn't interfere with hood/engine clearance. This approach allows the carburetor to remain virtually stock, since it is responding to vacuum instead of pressure. It is the most common of the available kits.

The simplicity of propane carburetion makes a

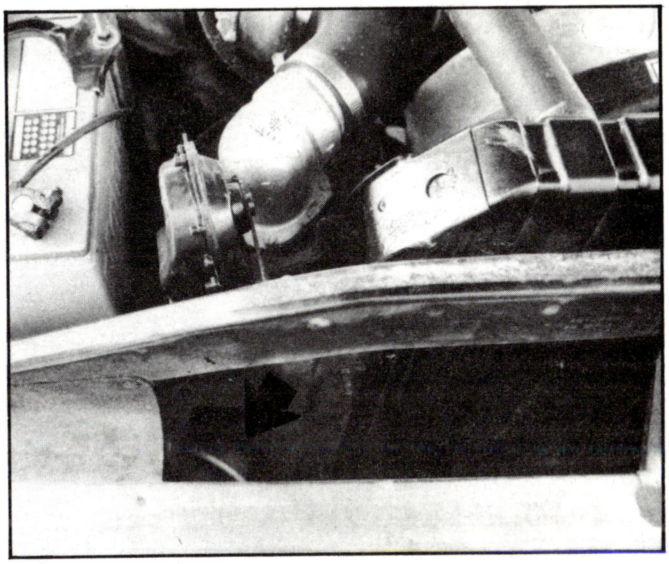

Since residual heat from the turbocharger will raise the induction temperature, it is especially important on any turbo installation to induct the coolest air possible to the turbo inlet. In this case an Ak Miller remote air cleaner was used and the air entry was extended through the radiator shroud (arrow).

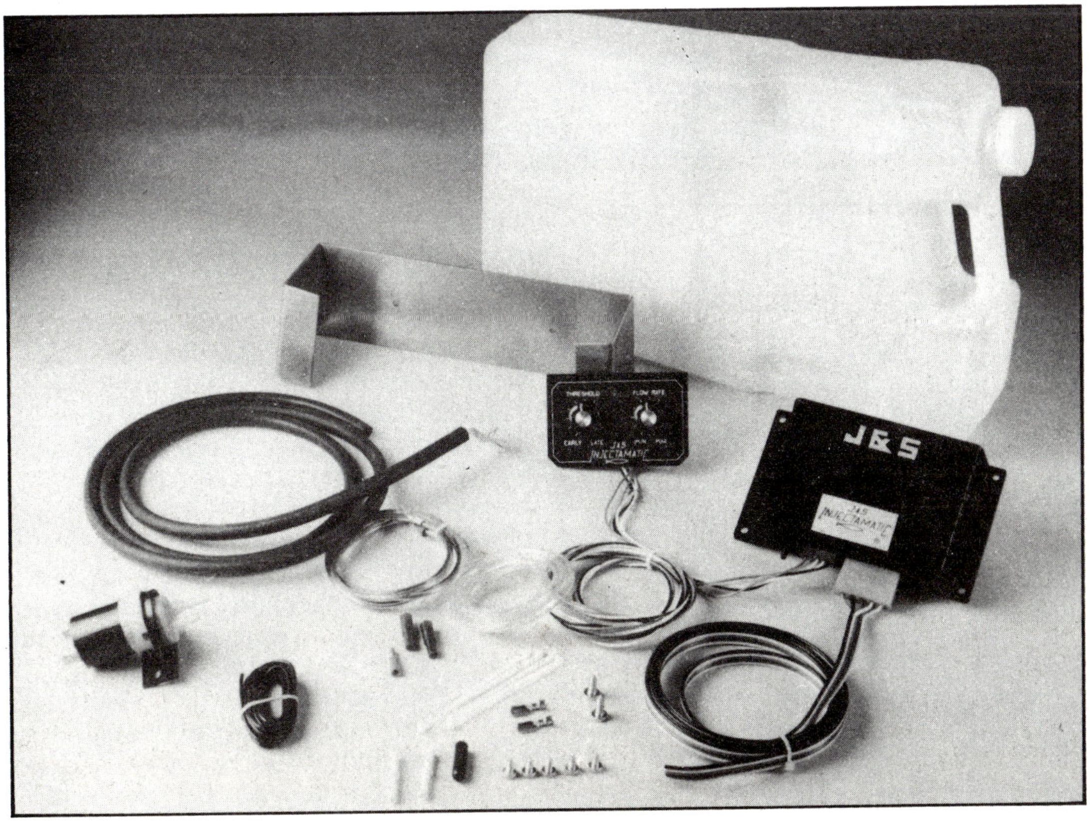

Water injection is a proven and reliable method for delaying the onset of detonation. There are many excellent electronically-controlled water injectors, such as this Injectamatic from J&S Electronics. It is adjustable from the dash for both threshold (onset of water flow) and rate of flow, and it has warning lights to indicate when the unit is operating and when the water level in the tank is low.

A steady supply of pressurized oil is essential to keep a high-speed turbo healthy, but a large and unrestricted oil-return line is just as important. The return line to the pan (which must be located above the standing oil level) must have an inside diameter of at least 5/8-inch.

world of difference when a turbocharger system is installed on a conventional engine. Because they have no floats, no fuel pump, and automatically compensate for different air densities, propane mixers adapt quite readily to either blow-through or draw-through turbo systems. With this being the case, the blow-through system becomes a much more attractive proposition. The propane mixer can be mounted on the intake manifold (in place of the stock gasoline carburetor) and all lines and linkages can be connected with relative ease. Pressurized air from the turbocharger is simply directed into the mixer.

No matter what type (or brand) of turbo system is selected, the operator must always remember that detonation is the major enemy of all turbocharged engines! Under boost conditions, a turbocharger pushes more fuel-air into the combustion chamber—in effect, raising the so-called dynamic compression ratio; i.e., since relatively more air-fuel is packed into the chamber during boost conditions, the effective ratio is actually higher than the static compression ratio). And even though propane is a comparatively high-octane fuel, everything possible must be done to counteract the additional heat created in the induction of a turbo system. This heat is the result of heat transfer (from the hot turbocharger housing) and the heat of compression. If this heat rise is not effectively reduced, the detonation threshold will be substantially lowered. As a result, the engine performance will be reduced—or in the worst case, engine damage may result!

The suitability of a specific engine for turbocharging isn't much different than it's general suitability for a propane conversion. In other words, the engine in question must be in sound mechanical shape with a

Sometimes there may be no easy way to route the air intake to the grille area. On this Cougar the air cleaner is mounted in the fenderwell and plumbed to the inlet with reinforced flexible hose.

The built-in, high-output coil from a late-model GM High-Energy ignition works well for an LP or turbo/LP engine. Removed from the distributor, it can be mounted on a fender panel and used on any engine, such as this Ford.

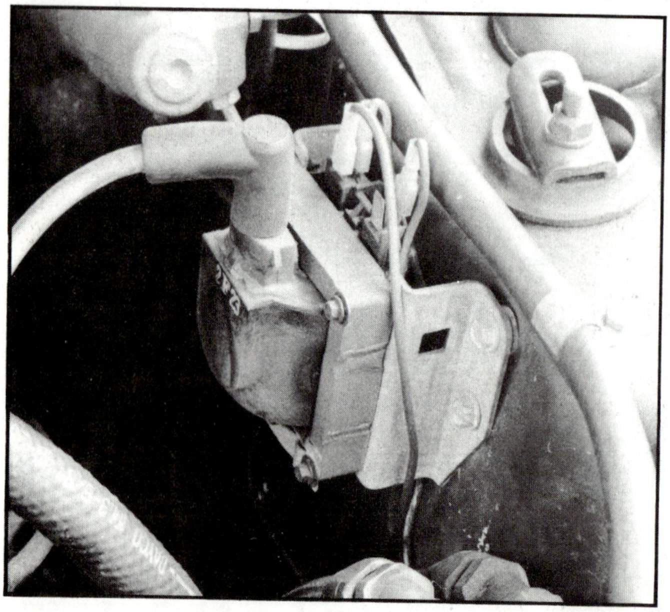

good ring seal, good valve seal and guides, and no vacuum leaks. Like propane, a turbo will not give new life to a tired engine, although it can definitely make for a more dramatic premature demise!

Most of the modifications for optimizing a gaseous-fueled engine (explained in other sections of this book) are still valid for a turbo-propane engine. However, some of the modifications that may be viewed as optional on a naturally-aspirated engine are absolutely essential in a turbocharged engine. For maximum performance and reliability under all operating conditions, a cold-air duct must be fabricated to channel cool air into the air cleaner. The heat risers in the intake manifold must be blocked to prevent unwanted heating of the intake mixture. The ignition distributor must be re-curved to match the new requirements of the turbocharged engine; and in most cases, colder spark plugs will be in order—one range colder than for a normally-aspirated propane engine. And, from a reliability standpoint, it would be well worth the effort and expense to install hardened valve seats in the cylinder heads.

A homogeneous intake mixture—particularly important in a turbocharged application (where you definitely don't want some cylinders to be leaner than others)—will also have some effect on intake-manifold design. Single-plane intake manifolds (designed primarily for high-performance applications), which have an open plenum chamber, work exceptionally well on turbo-propane engines. On standard gasoline, this type of manifold will improve engine output primarily at higher engine speeds; at lower speeds there is normally some loss of mixture velocity, which can lead to wet-fuel separation. But fuel separation is not a problem with propane, and the relatively straight runners of a typical aftermarket manifold promote good cylinder-to-cylinder distribution. On propane, this advantage is maintained even at lower velocities. This can be proven by observing the way the engine reacts to extremely lean mixtures. When a propane engine is fitted with an open-plenum manifold, like an Edelbrock or a Holley, the idle-mixture screw can be turned *all* the way in and the engine will still idle smoothly. The same engine won't perform like this with a stock, dual-plane manifold. (In fact, as Detroit has developed more sophisticated carburetion that allows leaner and leaner mixtures, future intake manifold designs may become more and more like "hot-rod" manifolds.) Therefore, when an engine is converted to propane, an aftermarket manifold will often prove beneficial in efficiency and performance—and even more so in a turbocharged setup.

As luck would have it, the best advance curve for a turbocharged engine is quite close to the ideal propane curve. From a practical standpoint, the most important consideration is the upper limit of the advance: the total should never exceed 32°. Long-time propane proponent Ak Miller has an interesting recommendation for finding the proper amount of initial advance. He likes to set the initial using a vacuum gauge on an idling engine. While watching the gauge, he advances or retards the distributor until the highest vacuum level is achieved. When you reach this point, back off 1/2-inch of mercury (indicated on the gauge) and lock the distributor in place.

If you use this method and then check the results with a conventional timing light, you may be shocked to see that the engine likes as much as 20° of initial timing! However, this setting should provide peak performance as long as the centrifugal-advance (provided by the distributor) is set to achieve full advance by 2500 rpm and total advance is limited to no more than 30°-32°. The vacuum-gauge technique practically ensures that the engine will provide maximum low-end torque (and we know already that a propane engine needs at least 10° more initial timing than a gasoline version of the same engine).

During full-throttle acceleration harmful detonation can set in so rapidly—and so disastrously—that backpedalling the throttle may not save the pistons! A proper tuneup will prevent detonation problems; but in fact, when the block is apart for a rebuild (for hard seats and a new set of rings), it would be more than worthwhile to install a set of forged pistons. (The Replacement Division of TRW, Inc., offers several low-compression, high-strength models that are highly respected by many turbocharging experts; these replacement pistons are available to fit most engines and can be obtained from all local TRW dealers.) Just make sure that the static compression does not exceed 8.0 or 8.5:1—actual, measured compression; not the "estimated" compression listed in sales literature.

Another simple modification is required when tur-

Eliminating the stock mechanical cooling fan yielded a 0.50 to 0.75 mpg improvement on both the Capri and Cougar. Two Flexalite electric fans were used, in place of the stock fan, to draw air through the radiator.

Of the several methods available to control boost, the intake controller is the simplest and least expensive. The TC-2 shown here is available from Ak Miller Enterprises. Installed in the plumbing to the intake manifold, the spring-loaded piston restricts the intake when the pre-set boost level is reached.

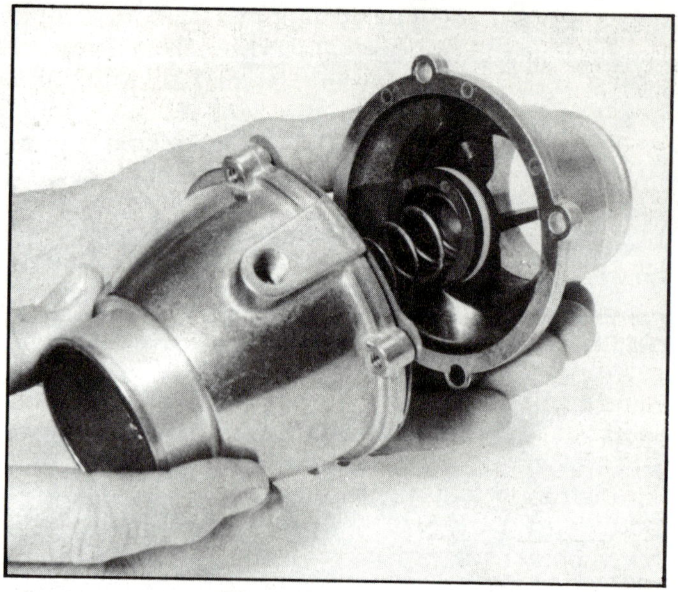

The spring inside the TC-2 can be changed to adjust the desired max-boost level. Replacement springs are available to allow a 5-pound, 7-pound, 9-pound, or 15-pound boost level.

bocharging. The turbo needs plenty of clean oil to withstand the high shaft speeds encountered by the spinning impellers, so pressurized engine oil must be supplied to the center bearing of the turbo. Fortunately, this is a relatively simple job, and suitable hardware is usually supplied with most turbo kits. In most cases, engine oil is tapped off the block (the access port for the oil-pressure sender is the most common place to accomplish this); and pressurized engine oil is directed through a small high-pressure feed hose to the turbo. This oil is then routed back to the oil pan through a separate return line, and the pan must be modified slightly to accept an appropriate hose fitting (a relatively simple job).

As the oil is delivered to the high-speed bearing, it encounters considerable heat; and as it exits the bearing housing, it has a tendency to foam and froth. For this reason, the oil-return line must be large in diameter—at least 5/8-inch inside diameter—and free of restricting bends. Any restriction in this line can cause the hot, foaming oil to fill the bearing case on the turbo, forcing oil by the seal and into the exhaust system.

If the engine in question suffers from inherent over-heating problems, it is highly recommended that a small, auxiliary oil cooler be added to the lubrication system. An oil cooler can reduce critical operating temperatures on any engine, but a turbo-propane powerplant has even more need of additional cooling—because of the extra heat created by the turbo. This small investment is well worth the cost!

As far as carburetion is concerned, no modifications are necessary for a turbocharged application. However, it is highly recommended that the final power adjustment (the power mixture setting, which is best finalized during dynamometer testing) should be set to 3% CO (instead of the usual 2.5% CO) to gain a slightly richer full-load mixture.

We have already discussed detonation problems, but the turbocharger adds a new dimension to detonation: too much boost. The turbocharger can increase the dynamic compression rapidly, and it's surprisingly easy to exceed the detonation threshold when the engine is operating under boost conditions. There are, however, three common methods of controlling the amount of maximum boost. The least expensive, but least efficient, is to control the boost through restrictions to the intake and exhaust systems. A stock muffler and stock air cleaner may be restrictive enough to limit maximum boost; but they will also limit the boost response, a symptom called *turbo lag*. This condition is most commonly manifested as a notable delay between the time when the throttle is pushed wide-open and the time when the engine actually receives boost.

Ak Miller Enterprises also offers a carburetor bonnet with a built-in TC-2 controller.

The boost controller must be installed somewhere between the compressor outlet and the intake manifold. A hose (arrow) connects the controller to intake-manifold vacuum/pressure, and as boost rises in the intake manifold, the manifold pressure pushes the restricting piston against the spring.

An exhaust wastegate is the most effective means for controlling boost. Unfortunately, a high-quality wastegate is expensive, but it limits boost—without over-stressing the turbo—by dumping exhaust pressure either outside the vehicle or back into the exhaust system (downstream of the turbo turbine). This Ak Miller Boost-Guard features simple click-stop adjustments to change the max-boost level.

A second method is to utilize a *boost controller* on the intake side of the system. Installed in the compressor pipe to the intake manifold, the controller is usually a spring-loaded assembly that constricts the induction (from the turbo compressor) when boost reaches a pre-set level. The level of constriction is adjusted by changing the spring pressure inside the boost controller.

These devices are inexpensive insurance for a tur-

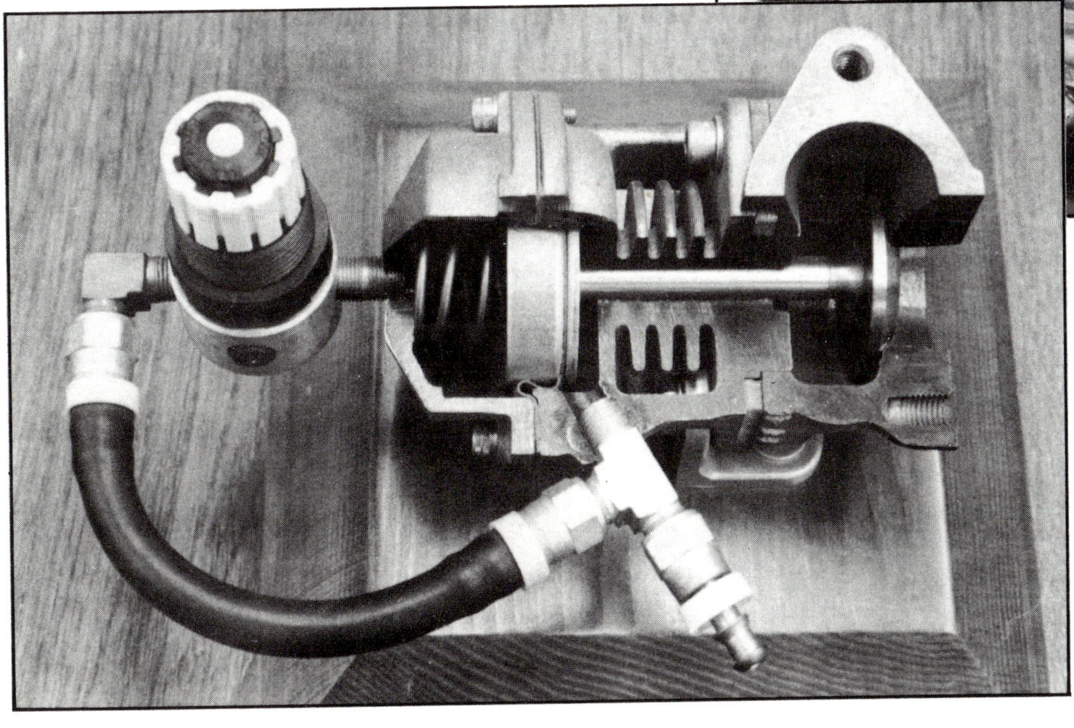

Like the turbocharger itself, an exhaust wastegate must be made of special materials to withstand the high temperatures. This cutaway shows how the pressure differential above and below the diaphragm causes the valve to open.

bocharged engine, but they do have a notable drawback. When the engine is operating near the boost limit, the controller constricts the output of the turbo and the engine feels like it's being strangled (which is exactly what is happening)! The throttling effect takes some getting used to but is apparent only when the engine reaches or approaches the maximum boost level. During normal driving and moderate acceleration, the restriction is not noticeable.

The optimum method for controlling boost is the most expensive (wouldn't you know it): a *waste-gate* on the exhaust side of the boost system. Since the flow of exhaust gasses control the turbocharger output, it would be natural to assume that if the exhaust flow is limited, the boost will likewise be limited. But the waste-gate limits the exhaust flow into the turbo not by constricting it but by channeling it around the turbo. When the engine reaches the maximum-boost level, the waste-gate simply allows excess exhaust flow to escape—either into the atmosphere or downstream into the exhaust system. This slows down the turbine but does not make the turbine work against a restriction (as a boost controller does).

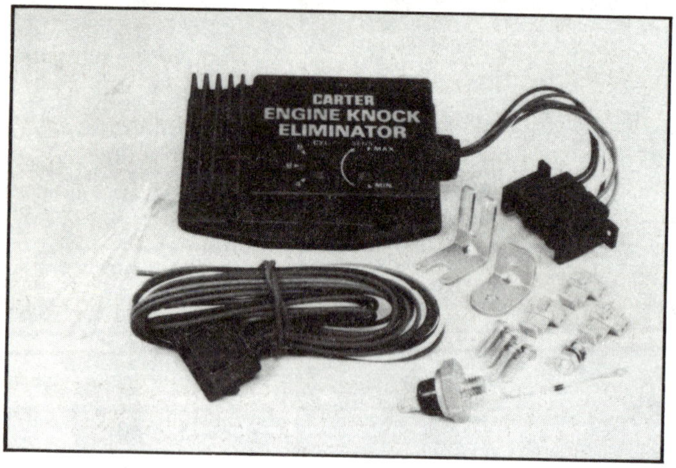

Though LPG has a relatively high octane rating, heat and pressure can produce detonation difficulties. Turbocharged engines on propane should have less tendency to detonate, up to boost levels of about 9 to 10psi (most gasoline engines are restricted to boost levels of about 6psi, unless fuel additives are used). However, propane/turbo engines can benefit from detonation countermeasures. One example is this ignition module from Carter. An engine-mounted sensor detects the telltale signs of detonation, and the black box retards the ignition until the detonation stops.

Since it is installed on the exhaust manifold and ahead of the turbine, the waste-gate has to withstand high heat levels and requires precise machine work with the best of materials—which makes it a reasonably expensive piece of hardware. The waste-gate usually has a simple bolt that is turned in or out to change the max-boost level. This is what championship racers (CART/USAC) fondly refer to as the "horsepower screw."

The safe level of boost, like other detonation-limited factors, depends on one main factor: the *workload* of the engine. In a passenger car, where loads are relatively light, the maximum boost on propane should be kept to about 9psi. A truck that must withstand heavier loads shouldn't have more than

Can a turbocharger be used on a dual-fuel vehicle? It sure can, and it works wonders in both modes. Ak Miller developed this 1980 2.3-liter Capri for Ford, and with catalytic converter in place and no internal engine modifications, it achieved 33mpg on gasoline and 22.5mpg on LPG! With both fuel systems filled, range was 1000 miles.

On the turbo dual-fuel Capri, the compressor blows into the propane carburetor, which blows directly into the chromed bonnet over the stock gasoline carburetor. Had the engine been suitably modified for straight LPG, propane mileage would have been much better (with some sacrifice of gasoline-fueled mileage). For best results, the engine should be built to maximize performance with the primary fuel, either gasoline or propane.

This Cougar XR-7, with stock 255 V-8, is another Ak Miller turbocharged dual-fuel project. Weighing 3800 pounds, this cruiser was a bit sluggish in stock form; but the conversion paid off with better economy and performance in both gas and LPG modes. With boost limited to seven pounds, the stock power went up from 85 horsepower (at the rear wheels) to 150!

about 5 psi boost.

Once the operator recovers from his initial enthusiasm, chances are very high that the engine will be operated 95% of the time without any boost at all. However, in a typical pickup truck with a camper (an excellent application for a turbocharged propane engine), the boost gauge may register only one or two pounds of boost while climbing a fair hill; but the truck will effortlessly climb the grade at 60 to 70 mph! And needless to say, this has made turbocharging and propane very popular with motorhome owners.

Another big plus is the benefit that turbocharging can provide for a dual-fuel conversion. A *turbocharged* dual-fuel setup loses less efficiency and will probably have better performance in either mode than it did before the conversion. The intake restrictions in the carburetion becomes less of a problem when the turbo is blowing through them. For example, Ak Miller has built several turbocharged dual-fuel engines; and during some development tests for Ford Motor Company, some of his turbo-propane vehicles were faster in quarter-mile acceleration tests than in the stock mode—and the turbo-propane mode was faster yet than the turbo-gasoline mode!

If we were to give you one caution regarding turbocharging on dual-fuel installations, it would be to watch for restrictions in the electric lock-offs. Some of these lock-offs (the ones for the propane side of the system) have holes that are too small for a turbocharged application, though they work well enough on a normally aspirated engine. The turbocharged engine in the boost mode requires more fuel delivery than under normal circumstances, and your propane carburetion can supply the correct ratio of fuel to air. But the lock-off may prove restrictive to the propane flow if the hole through the middle isn't at least 3/16-inch or 1/4-inch.

The turbocharged dual-fuel setup on this 454 Chevy truck engine required two vaporizers to supply the large mixer (when in the propane mode).

CHAPTER 6
TROUBLE SHOOTING

In most cases troubleshooting a propane engine is easier than a comparable gasoline vehicle because the simple nature of gaseous-fuel carburetion eliminates many common difficulties. When the carburetion does prove to be at fault, the checks and repairs are straightforward.

Unlike gasoline carburetion, propane mixers are insensitive to idle-mixture adjustments. Most engines idle smoothly no matter what the setting, so only an exhaust-gas analyzer will tell the true story.

MAINTENANCE & TROUBLESHOOTING

Probably the most important advice we can give you about troubleshooting a propane-fueled engine is to leave the propane system alone. Most drivers or mechanics will be quick to blame the fuel or carburetor for any problems that arise; but nine times out of ten, the answer will lie somewhere else. *The ignition system should be the primary suspect at all times!* In the engine-building chapter, we went into detail about the ignition requirements of a propane engine; if the system is updated to these specs and the advance curve is altered to suit the requirements of propane, there should be no trouble with the ignition.

If you have a late-model car with electronic ignition, you shouldn't have to worry about checking point gap; and the plugs should last about 30,000 miles if they are cleaned and re-gapped at the 15,000-mile interval. Right after a conversion, perhaps at the first 5,000 miles, is a good time to check the plugs. Use a good magnifying glass to examine the insulators for indications of heat. If the center insulator is a dark grey color, the plug range is probably too cold; and if the center insulator is blistered, the range is too hot. When the center insulator is only slightly colored—grey or light brown—and there are no blisters, the plug heat range is in the ballpark.

Remember, though, that unless you have the timing and advance curve properly set up for propane, you're wasting your time looking at the plugs. A plug of the correct heat range will usually look overheated if the advance curve is stock (a stock curve pulls in the advance too slowly).

And, if the engine has been broken in properly (on gasoline), there should not be any signs of oil-burning or soot on the plugs. On propane, you virtually cannot run the mixture adjustment rich enough to cause carbon deposits on the plugs (such as a rich gasoline engine will); so any soot on the plugs will be a result of oil burning. When crankcase oil escapes into the chamber, it will show up on the spark plugs first on the rim of the plug shell, later inside the shell, and finally on the bottom of the porcelain. If the engine is burning oil, don't blame it on the propane; perform a leak-down test and find out where the trouble is. (If you shut the engine down after it has reached operating temperature and a puff of smoke comes out the tailpipe when the engine is fired up about a half-hour later, there is a good chance that the exhaust valve guides are leaking.)

While still in the realm of ignition problems, it is important to remember that occasionally a starting resistor will fail, causing a voltage drop during the critical cranking-over period. You can check this easily with a voltmeter. The starting resistor is designed to allow full battery voltage to reach the coil during starting and to cut back on voltage after the engine is running. Check the available voltage at the battery with the voltmeter, then compare with a reading taken at the coil; there should be no appreciable drop when the starter motor is engaged.

Any good auto-repair manual should give you additional guidance on troubleshooting a conventional ignition system. However, if the engine has an electronic ignition (and these miracle "black boxes" *can* fail), you may need a factory shop manual to adequately check out the entire system (nothing more complicated than a 12-volt test light and a volt-ohmmeter are required for most test procedures).

Nonetheless, even with the best of ignition systems, the distributor may malfunction or fail. In the past, this was not a common occurrence; but in recent times, distributor problems have become more commonplace—usually because of a lack of lubrication in the advance mechanism. Without lubrication, the advance weights and pivot pins will rust. This is primarily caused by condensed moisture inside the distributor housing; and once the internal components become oxidized, it's only a matter of time before the advance mechanism becomes worn and sloppy, or worse, frozen solid. We've also seen some late-model GM distributors with pivot pins that had become so rusty that the advance weights sheared the pins at high speed and flew out of the distributor (right through the side of the cap).

When the distributor is torn down to modify the advance curve, make sure all the weights, pins, and plates are clean of rust and dirt; and lubricate these components with small dabs of distributor cam grease (do not use any other type of lube on these parts). And remember, even if the distributor doesn't pitch parts around the engine compartment, a rusted advance mechanism can still be potentially dangerous on propane. If the mechanism sticks in the fully advanced position, the engine will detonate at low speeds and this will likely cause a blown head gasket.

The distributor centershaft and cams can also wear down (on points-type ignitions), causing erratic ignition advance. When the distributor is apart for maintenance or modification, check the centershaft clearance (side-to-side play in the bushings, not up-and-down play). Too much free play will cause erratic timing and may allow the rotor tip to rotate in an elliptical (instead of round) pattern. This will cause timing irregularities; and in extreme cases, it may actually allow the rotor to hit some of the secondary terminals inside the distributor cap. (When some of the terminals inside the cap look chewed up and others don't, it's a sure sign of a sloppy shaft.)

With points-type ignitions, the cam that operates the points can wear away, usually from the lack of proper lubrication on the rubbing block (the rubbing block and cam must be lubricated periodically with distributor cam grease). To check this, find the two cylinders that are at TDC at the same time. (On inline engines, these are usually the first and last cylinders.) With the engine running, use an accurate timing light on both of these cylinders; if the timing indications are different, then chances are good the points cam is worn.

PROPANE TROUBLESHOOTING

It won't happen very often, but occasionally things can go wrong with the propane components. However, the great beauty of a propane system is that *the few things that can go wrong are all easily corrected.*

The basic fuel system is composed of a tank, high-pressure line, a filter/fuel-lock, a vaporizer/regulator, and a mixer. Starting at the rear, let's examine the tank itself. First, determine if the tank does indeed have propane in it. This sounds simple when there's a gauge right on the tank, but the floats for these gauges have been known to stick in the *up* position and give a false reading. Open the outage valve to see if any vapors come out.

If the tank *is* empty and you're not at a propane filling station, you'll need an auxiliary tank to get going again. If the installation is straight propane, it's a good idea to keep a small spare tank (like a 5-gallon camper bottle) at home for emergencies. With such a tank is is possible to unscrew the high-pressure line from the main tank (after closing the valve, of course) and

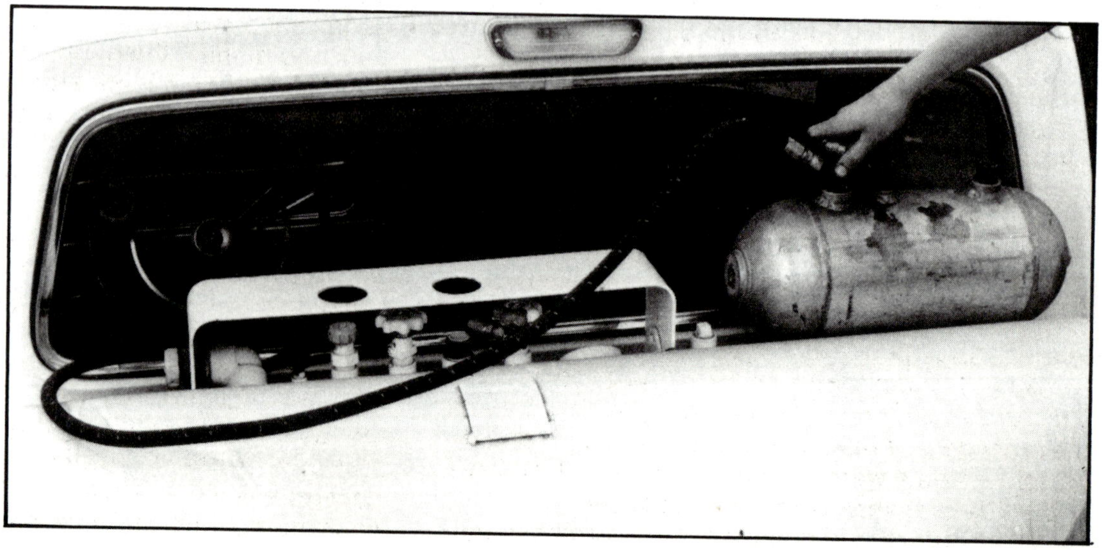

It is convenient to have a small auxiliary fuel tank for the rare occasion (hopefully) when inadequate planning leads to an empty fuel tank. A short hose is also required to connect the auxiliary tank to the vapor valve of the main tank.

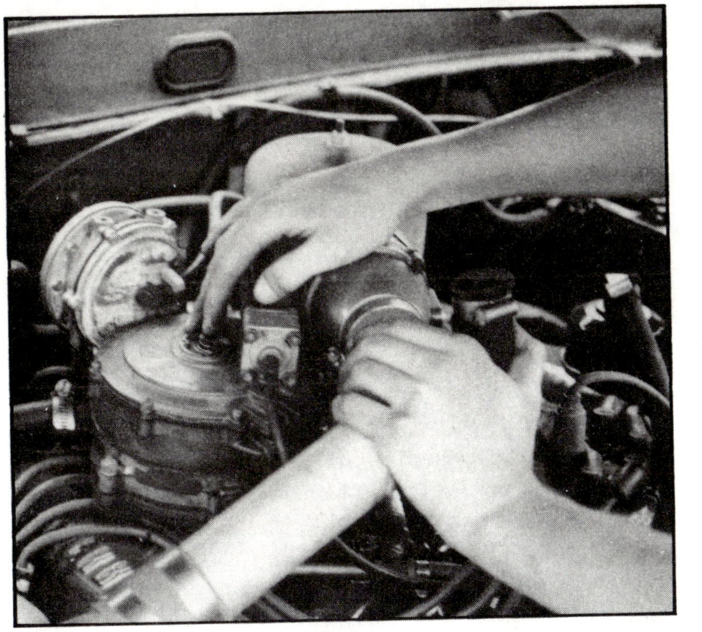

To determine if there is vapor available from the converter, disconnect the vapor hose (arrow) from the carb and depress the primer button on the vaporizer. There should be an audible sound when the vapors are released. After one or two depressions of the primer, vapor release should stop; the converter needs at least a 1.5-inch signal (WC) before it will begin delivery.

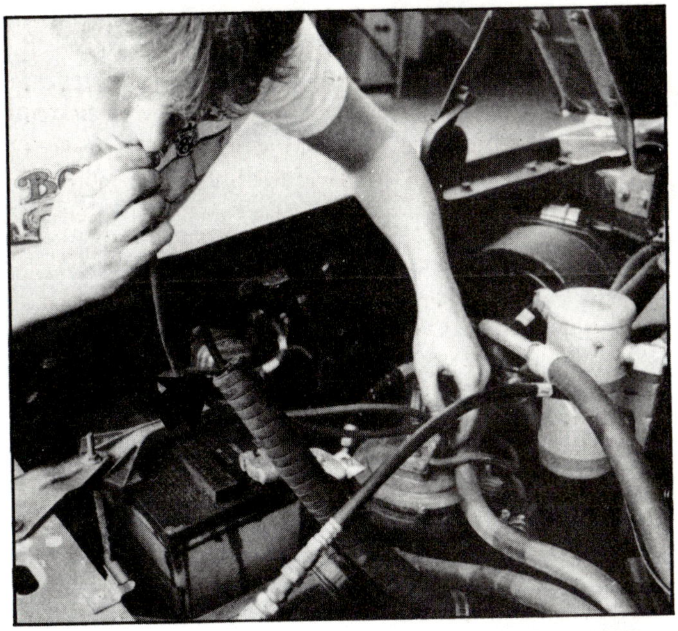

To make certain that the filter/fuel-lock is working properly, leave the vapor hose (arrow) disconnected from the mixer and apply a suction on the vacuum line from the fuel-lock (have an assistant turn the ignition key to activate the electrical lock-offs). When suction is applied to this line, fuel should flow from the fuel-lock to converter. The flow should stop when the suction is discontinued.

couple it to the emergency tank. Then, you can open the valve on the smaller tank and be on your way. (Note: Always be certain to wear gloves and goggles whenever disconnecting the high-pressure line from a tank, the extremely-cold propane in the line can cause severe burns if it contacts your skin in liquid form!)

Assuming that there is propane in the tank, check that the excess-flow valve hasn't shut off the supply at the tank. Remember, whenever you open the main valve on the tank (some operators like to shut the main valve when the vehicle is parked overnight, which isn't really necessary for safety purposes but does create an effective anti-theft system) just crack the valve open at first, then open it *slowly* all the way. When you open the valve too quickly, the tank tries to fill the entire system very rapidly and this can cause the excess-flow valve in the tank to shut off!

Next, proceeding from the tank, check to see if there are any problems with the high-pressure line. Is it dangerously chafed because it isn't firmly attached to the chassis at strategic points or is it too close to a hot exhaust system? If the latter is true, you may find that the hydrostatic-relief valve may pop when the car is put away hot and the tank valve is closed for the night. Rising pressure in the high-pressure line can pop the valve open, but it will only be for a split-second, until the valve reduces the pressure and cools off the line. And speaking of relief valves opening, don't fill the propane tank full in cold weather and then park the vehicle inside a warm garage for the night. This may cause the pressure-relief valve to blow.

When either the tank valve or the hydrostatic valve blows open to relieve pressure, it's not normally a serious occurrence, but it can be a matter of concern if it happens indoors and propane vapors are released inside a building. Being heavier than air, the propane will settle to the floor, and in fleet operations, where several vehicles may pop reliefs (because of the above), it is possible for the vapors to accumulate and create a potential fire hazard (such accumulations can be spontaneously ignited by a pilot light on a water heater or stove).

Now we come to some of the propane components that are underhood. Ideally, whenever you plan to work on any of the propane components ahead of the tank, the high-pressure line should be purged to avoid getting burns from propane spattering when the fittings are loosened. This is accomplished most easily by closing the valve at the tank and running the engine until it dies from lack of fuel. This will completely purge the high-pressure line, and the other components should also be free of high-pressure liquid.

Before any of the components are removed or disassembled, make a few simple checks with the system still full of liquid fuel. Remove the large-diameter vapor hose (at the carburetor end) that goes from the converter to the carburetor. With the engine off, there should be no leakage of vapors here. Press down on the primer button of the converter, and there should be a slight rush of vapor from the hose. If the filter/fuel-lock is functioning properly, the vapor should stop as soon as the line between the fuel-lock and the converter is emptied. Now pull off the small vacuum hose that connects to the vacuum fuel-lock (or turn on the ignition key momentarily if the system has an electric lock-off) and apply a light suction on the hose end. This should permit fuel to flow through the converter and delivery hose, and this flow should stop as soon as the suction on the vacuum line is stopped. This indicates that the fuel-lock is working properly.

If the filter/fuel-lock doesn't work properly in the above tests, disassemble it, clean it in kerosene, and install a repair kit. There aren't very many parts inside, so the repair-kit installation can normally be accomplished without elaborate tools or specialized knowledge. Use the manufacturer's repair manual or kit instructions; most propane carburetion repairs can be accomplished with nothing more complicated than a screwdriver.

The other side of the filter/fuel-lock is the liquid propane filter, which should not give any trouble in 100,000 miles of operation. The filter pad collects impurities (rust particles and welding slag) that may come out of a new tank, and also collects any large particles of dirt that may be in the propane fuel (commercial-grade propane usually contains some dirt and oils from the large compressors used to pump it through storage facilities). The filter pad can be removed for cleaning, but this will rarely be necessary.

The converter will seldom cause any trouble if it has been installed properly. First, however, be certain that the converter *has been installed properly.* Is the converter mounted with the vapor output line pointing slightly downhill? This is important so that impurities (slight amount of oil) enter the engine and won't collect in the converter, but will flow along to the carburetor. Is the converter mounted in such a way that it is the highest point in the cooling system? If so, you may get an airlock in the cooling

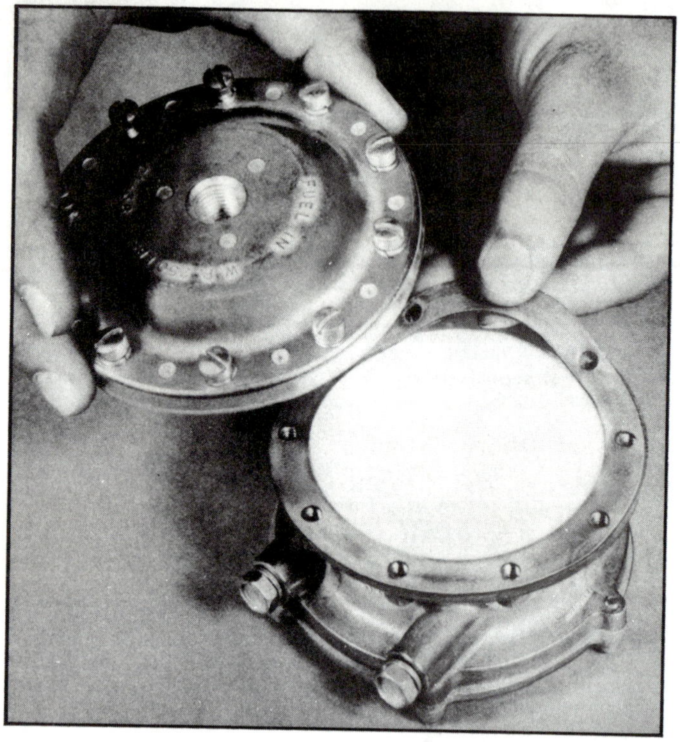

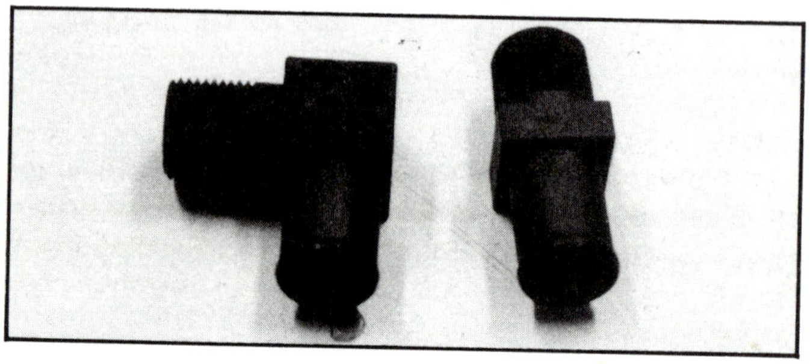

To prevent electrolysis from eroding the converter, never use steel water fittings. Use conventional brass fittings or these plastic fittings from Impco. Plastic fittings will not cause cracks in the converter if they are overtightened.

The vacuum filter/fuel-lock is very simple—just a diaphragm and a filter pad. It can be completely disassembled and rebuilt with a screwdriver.

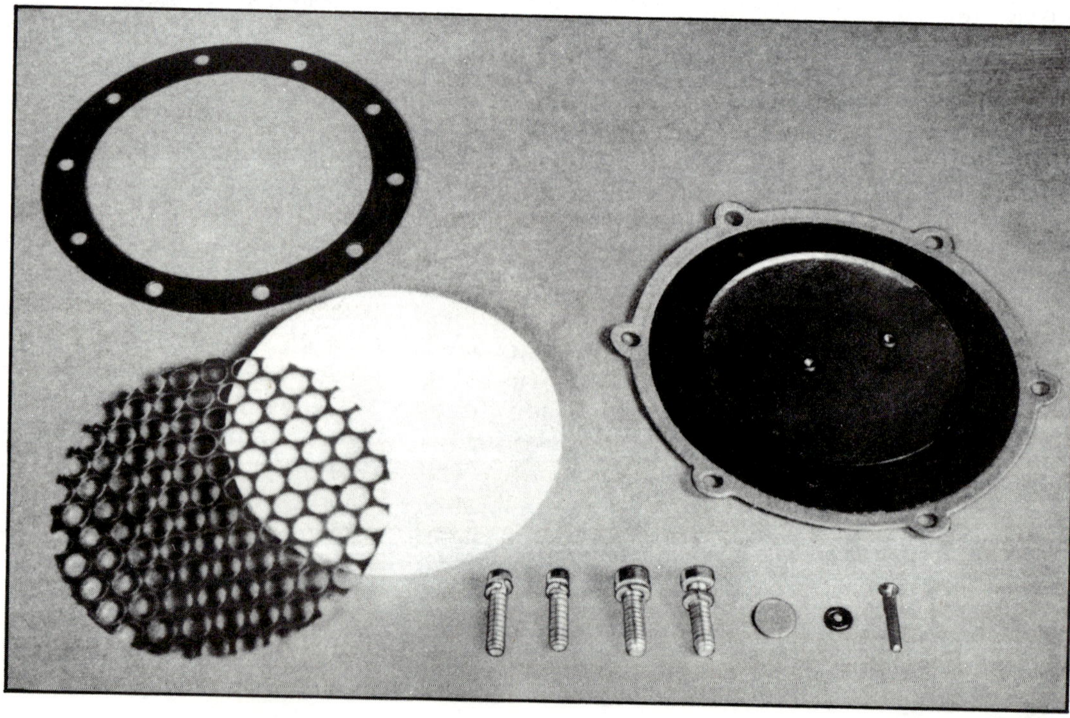

A repair kit for a filter/fuel-lock consists of a new diaphragm (right) and filter pad. Here, beneath the diaphragm, is a new Viton valve-spring seat and O-ring. Most manufacturers supply complete step-by-step instruction with rebuild kits.

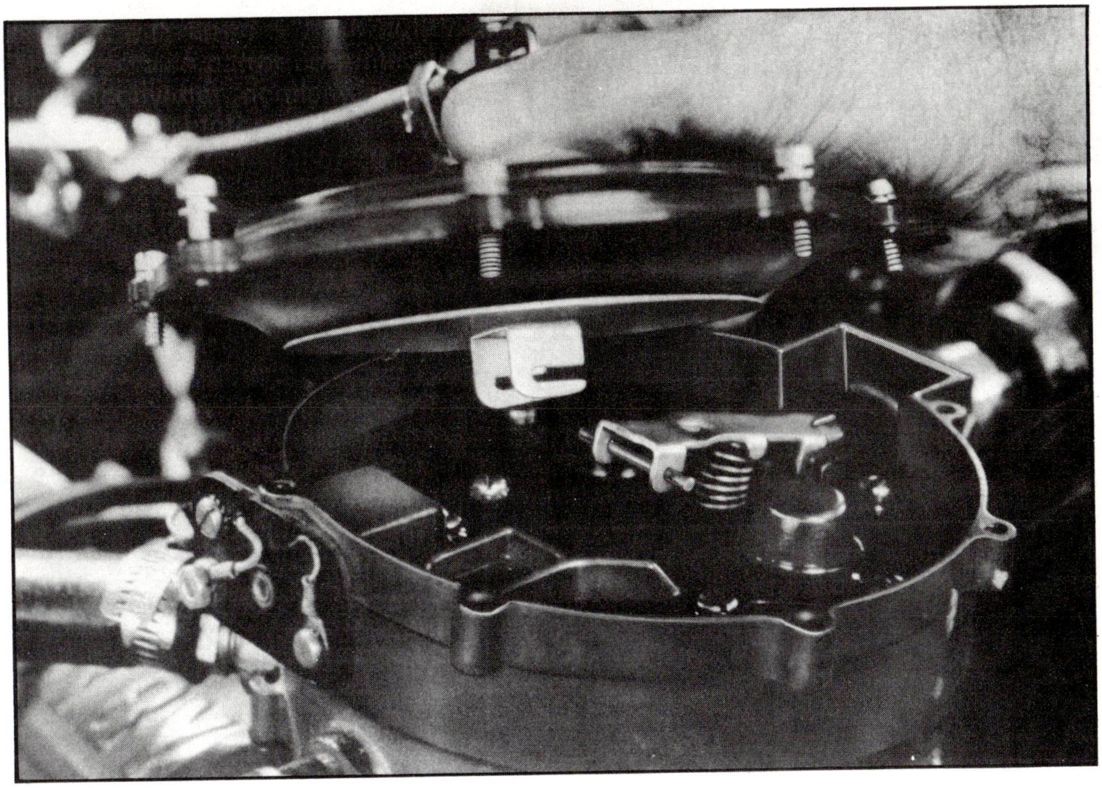

Removing the eight screws from the cover of a converter reveals the diaphragm and the slotted metal tabs that fit over the pin on the secondary regulator lever. A converter can generally be overhauled while still in place on the engine, even with the fuel and water hoses in place.

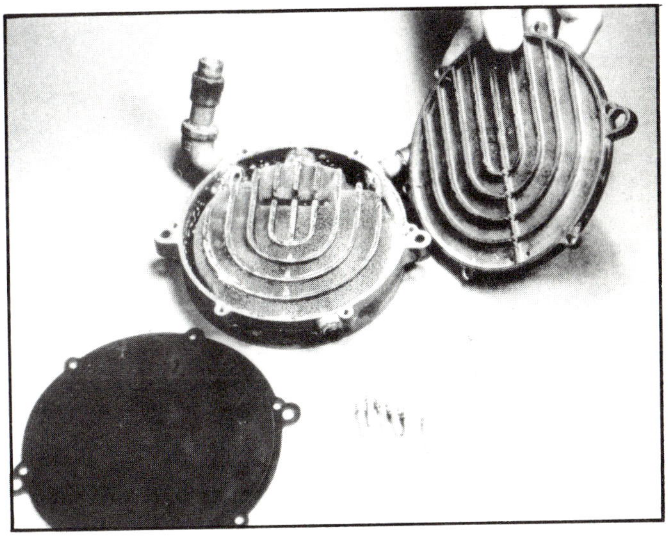

Most converter problems are related to the condition of the cooling system. This is the bottom portion of a converter from a vehicle with a badly maintained system. When the heat exchanger is clogged with rust and scale, the converter may "frost up" during operation. When flushing a cooling system, always disconnect the water lines to the converter to prevent loose rust from settling in the converter chamber.

system that will cause overheating. Are the water lines connected properly (flow both in and out) and are the hoses in good shape? Are the water fittings on the converter made of brass or plastic? They should be, because steel fittings screwed into the aluminum casting will cause electrolysis that will eat away the converter casting over a period of time. Another possible problem is that the fittings have been lubed with Teflon tape or sealant and have cracked the housing. The slippery Teflon makes the pipe-thread fittings go in so easily that some mechanics have a tendency to overtighten them.

A converter should provide trouble-free operation for at least 50,000 miles, and in most cases it will last much longer before any attention is required. In the unlikely event that a malfunction does occur, the trouble can usually be traced to the Viton seat in the regulator portion of the converter. This seat can become worn if dirt particles get past the filter and fall out of suspension as the liquid propane is vaporized. When the seat as been damaged, you'll know right away! This allows high pressure to build in the converter, causing engine flooding and an exceedingly rich mixture that cannot be corrected with the idle screw. The engine will also be very hard to start when it is hot-soaked. Another dead giveaway is when a cold-soaked engine fires up the next morning *instantly*, without the characteristic delay associated with propane fuel systems. This indicated that leakage past the seat in the converter has already filled the manifold with fuel.

This problem can be corrected with a rebuilding kit for the converter (normally available from the converter manufacturer or a local dealer). This kit will usually contain new diaphragms, Viton seats, and instructions for the proper disassembly and reassembly of the unit.

Other than Viton seat wear, there are only two other factors that may require the disassembly of the converter. First, if the cooling system hasn't been maintained and is full of rust and scale, these deposits may collect in the bottom chamber of the converter. This condition is indicated if the converter frosts up or freezes during normal operation. And, even if the cooling system has been maintained regularly, someone may have "flushed" the system with an acid-type cleaner that dislodged scale from the radiator *but deposited the residue in the bottom chamber of the converter*, where it restricts the cooling system as well as limiting the ability of the converter to vaporize the liquid propane.

It's very important in any automotive engine to

Manufacturers specify the correct pin height for the secondary regulator, which can be checked by holding a straight-edge across the top of the regulator body.

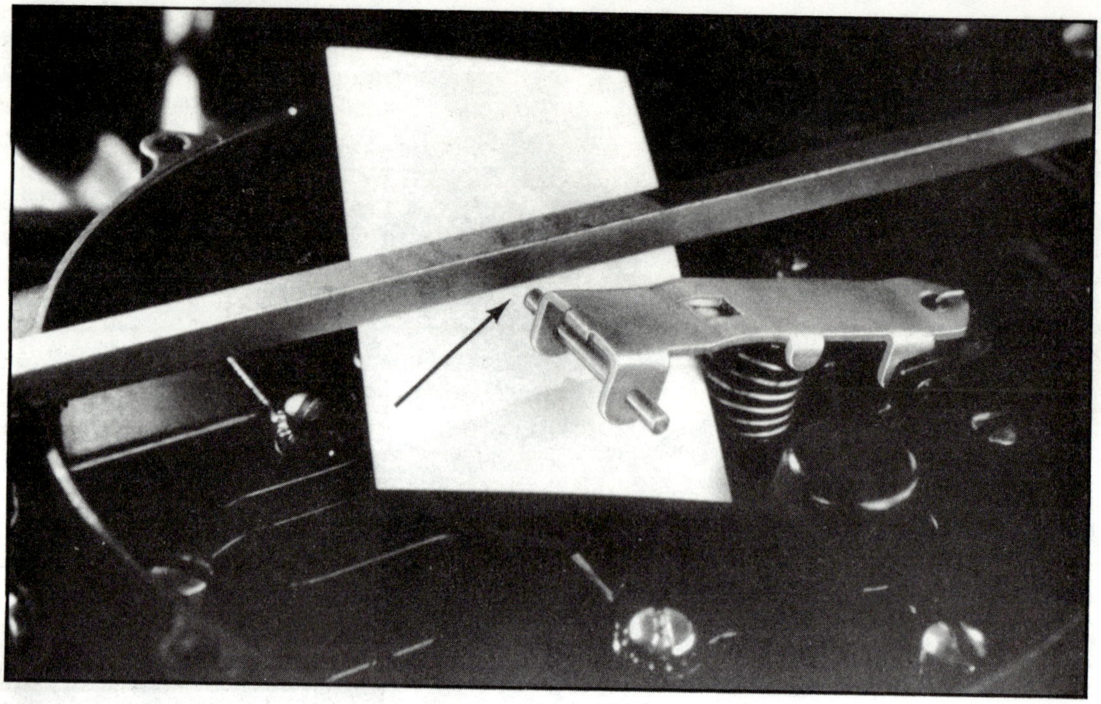

The primary valve is under the regulator body. During a rebuild the valve is generally replaced. This shouldn't be necessary until at least 50,000 to 100,000 miles of operation.

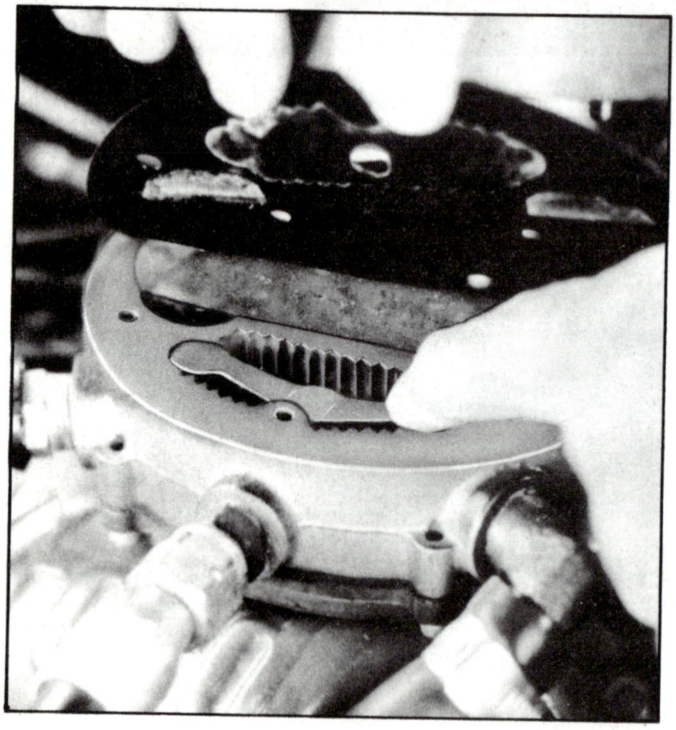

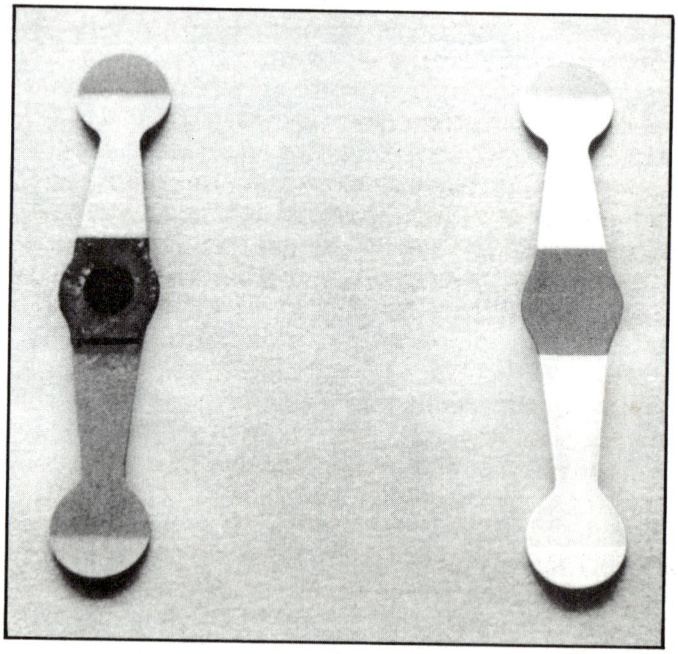

At left is an old primary valve seat that has been eroded from the sulphur deposits in the fuel. At right is a new seat.

keep the proper concentration of anti-freeze in the cooling system. In general, this prevents corrosion problems and increases cooling efficiency. Most engines need all of the cooling capacity possible—especially in the summer and in warmer climes—but this is particularly critical with a propane engine because adequate coolant flow through the converter is essential for proper converter operation. And if the cooling system is ever flushed with cleaner, the coolant circuit should be temporarily routed around the converter by disconnecting the two water hoses from the converter and joining them together with a short length of tubing and two hose clamps.

If the converter must be disassembled to install a repair kit, the bottom chamber—where the coolant circulates—can be cleaned with a hand-held wire brush. The rest of the converter parts can be cleaned with standard solvent. (We hope you know better than to clean any parts with dangerously flammable gasoline). And, if the unit in question is an Impco converter, under no circumstances should it ever be cleaned in a "hot-tank." The water chamber of the Impco converter is impregnated with a special coating to protect the aluminum from corrosion damage and to aid heat transfer; the caustic solution in the hot tank will damage this coating!

Before a new or used converter is installed it is always a good idea to check for fuel leaks. Feed a pressurized air supply into the fuel inlet of the converter and flip the low-pressure lever quickly four or five times. This should fill the chamber with about 10 psi of air pressure and the converter should hold it. Let the converter set for a few minutes and then push the lever again, this should release the air. If the air pressure has dissipated before the lever is opened, the seat or the diaphragm is leaking.

Finally, if you suspect the carburetion problems may lie in the propane mixer, the first step should be to measure the gas inlet pressure. This can be done only with a *water* manometer. You cannot use a gauge that reads in inches of *mercury*! Propane carburetion equipment functions on very small signals compared to gasoline carburetors. (Remember that one psi of pressure will move a mercury manometer 2.04 inches and that same pressure will move a water column 27.72 inches.) The vapor inlet pressure for Impco mixers should be 1.5 inches of water *negative*. Other venturi-type mixers are usually set to deliver gas at 0.1 to 0.5 inches *negative*.

A water manometer can easily be constructed by forming some clear plastic tubing into a U-shape and mounting it on a piece of plywood with a ruler attached next to the bend (see illustration). Fill the manometer half-way with water (colored water is easier to read, use a little food coloring), plumb the manometer to your vapor hose as shown and measure with the ruler how much the water moves on one side of the U when pressure is applied to the other end during cranking and idle.

If there still seems to be problem, even though the vaporizer is delivering gas at the proper pressure, then take off the carburetor cover and examine the diaphragm. It should be quite flexible (they have been known to last for five years or more), and without any pinholes or charring (caused by backfires). For those applications that function under high heat (such as turbocharged engines operating as blow-throughs or tight underhood areas like vans, motorhomes, etc.), Impco offers special silicone/fiberglass diaphragms.

Assuming that the diaphragm is okay, inspect the air valve cup and the gas jet. Vibrations may have caused the guide fins on the jet to groove the cup. With the carb reassembled, and off the manifold, reach up through the throttle body with a round-end rod and check that the spring closes the valve—lift it up

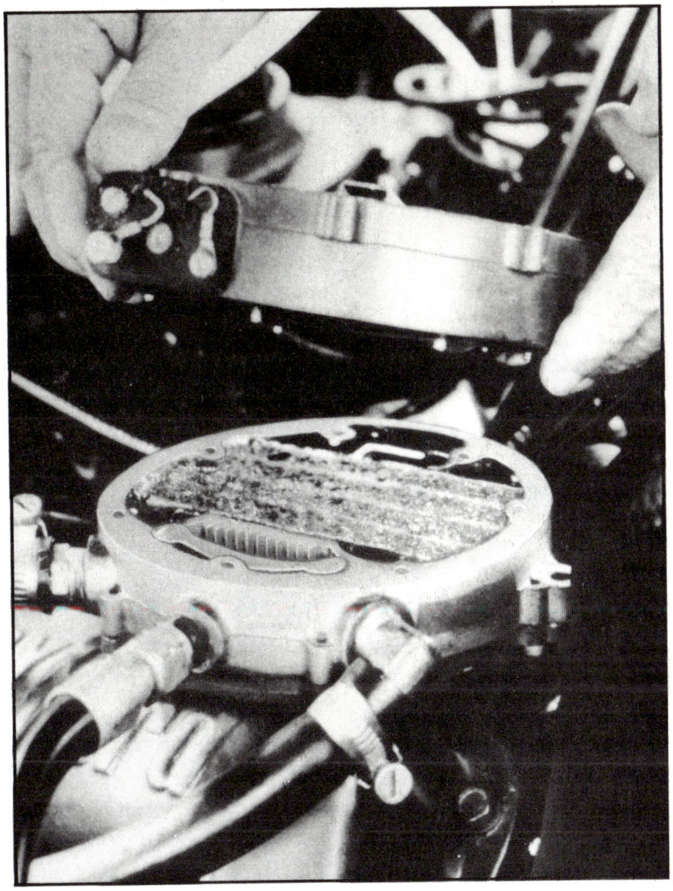

Here, behind the primary valve, is the converter sponge, which absorbs liquid fuel during cold starts. It works quite effectively regardless of age or mileage.

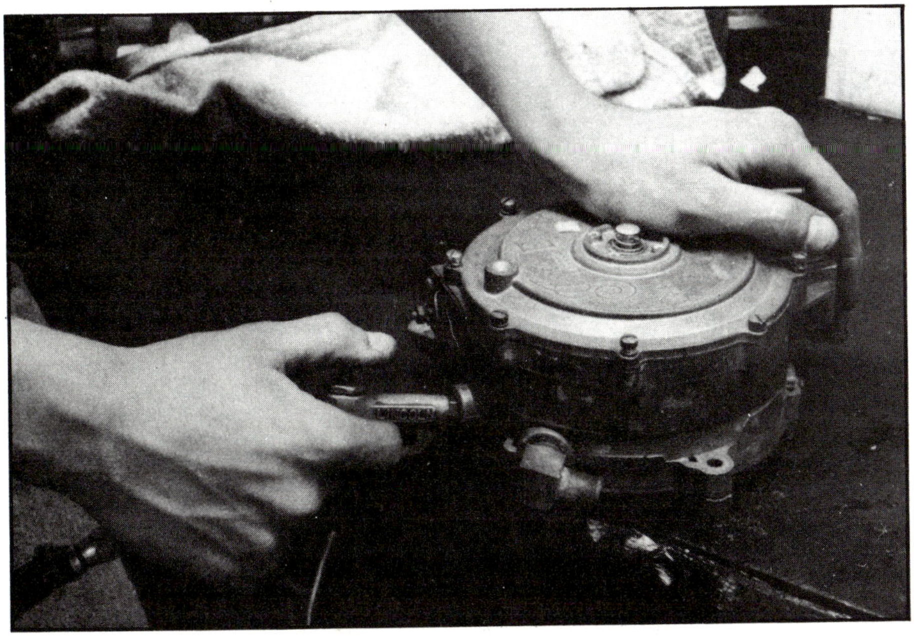

To determine if the converter pressure-regulating valves are working, apply pressurized air to the fuel inlet. The text describes this checking procedure in detail.

several times with the rod and make certain the spring pulls it fully closed. To make sure that there isn't a leak, suck on the gas inlet when the valve is closed; you shouldn't be able to draw air through it.

Rebuilding of the propane components will seldom be required under normal circumstances, but *it will always be required when the system has been removed from service and stored for any length of time.* This occurs most often when the operator removes the equipment to sell a vehicle and a few months later installs it on another vehicle. Under normal operating conditions the diaphragms and seats in the converter and mixer are coated with a light film of oil (from the engine vapors and oils that come out of suspension from the vaporized propane), and when put in storage they will dry out and develop cracks. This, of course, doesn't apply to new equipment that has never been exposed to oils.

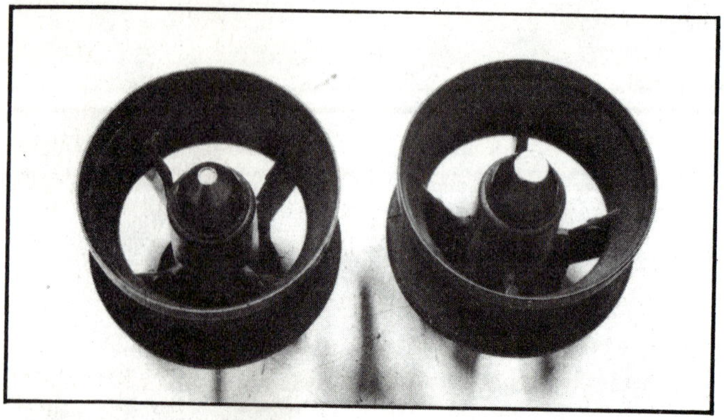

The gas jets have different shapes, depending on the application for which the mixer is designed. Some are also grooved. During a rebuild, both the jet and the seat should be checked for signs of vibration wear (a prevalent problem on small, in-line engines).

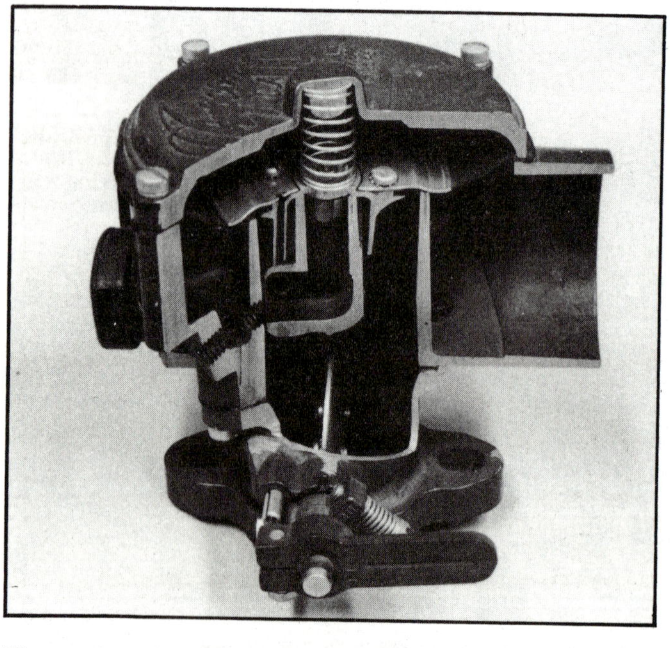

Those who have experience with complicated gasoline carburetors—which can be crippled by a tiny particle of fuel-born dirt—will appreciate the wonderful simplicity of a propane carburetor. It has one moving part: the air valve.

A restricted or dirty air cleaner can also create mixture problems that will reduce fuel efficiency greatly. Keep the filter clean, and if the filter size is restricted because of space limitations, it will be necessary to install a *balance line* between the converter and the carburetor. Route a 3/8-inch hose from the atmospheric vent on the converter (unscrew the little brass screen) to the balance port on the carburetor. This will keep the mixture reasonably balanced, even if the air cleaner becomes restricted through use. If backfiring occurs after the balance line is installed, the line is too restrictive and is causing a lean condition. Make sure the hose is at least 3/8-inch (inside diameter) and use straight (not angled) fittings at both the converter and carburetor.

If there is a drawback to the air-valve system used in gaseous carburetors, it is the high sensitivity these carburetors have for vacuum leaks in the manifold. If the intake system is inadequately sealed, the engine may be hard to start because there isn't enough vacuum to lift the diaphragm. In most cases the culprit will be a bad carburetor or intake-manifold gasket seal, but a low manifold signal can also be caused by a stuck PCV valve, worn valve guides or bad valve seating. However, an engine in relatively good condition should generate more than enough vacuum during engine starting for most propane carburetors. If you have a vacuum gauge (typically calibrated in inches of mercury) in the car, the needle should "dance" at least an inch of vacuum during cranking. Remember that the carburetor only needs 1.5 inches of water, and the inch on the common mercury gauge represents 13.6 inches of water.

When the air valve is seated properly in the gas mixer, a suction on the fuel-vapor inlet should not produce any airflow through the mixer.

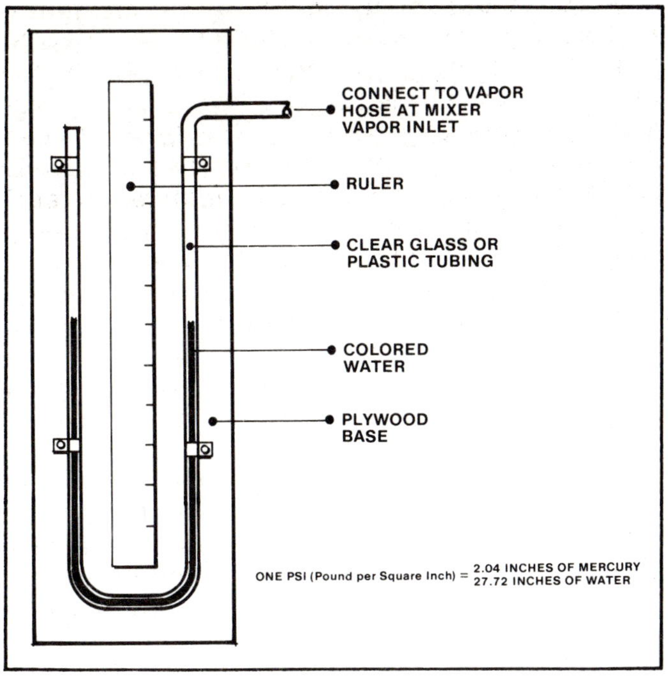

It is possible that an engine with very low manifold vacuum (due, most likely, to leaks) won't have enough signal at the mixer to draw the vapors from the converter. This is easily checked with a homemade water manometer.

The insulator gasket used under some gasoline carburetors (common on GM cars) is a potential source for vacuum leaks. A backfire blew this Quadrajet gasket apart. If you must reuse an old insulator gasket, remove the bushings (around the stud holes), reduce the thickness of the bushings with a file, then put them back into the gasket. This allows the gasket to be clinched more tightly between the carb and the manifold, reducing the chances of vacuum leakage at the base of the carburetor.

If you suspect that a low vacuum may be caused by leaky intake or carburetor gaskets, don't try the old mechanic's trick of pouring gasoline along the manifold/head interface to see if the idle picks up. Gasoline is extremely dangerous! A much neater and safer method is to use an unlit propane torch (typical home handyman torch with small throwaway fuel can). Turn on the torch slightly—without lighting it—and pass the torch around suspected areas while the engine is idling. When you get to an area where there is a vacuum leak, the idle should pick up noticeably due to the enrichened mixture!

The thick insulator gasket used under many OEM gasoline carburetors is a common source for vacuum leaks. The problem is greatest with those insulator gaskets that have plastic bushing inserts where the carburetor hold-down bolts or studs pass through the gasket. These bushings prevent the gasket from being crushed excessively when the carburetor hold-down fasteners are tightened, but the problem comes on the *second* use, after the gasket has already been compressed once. The plastic bushings now limit the amount of gasket crush and the usual result is a vacuum leak that cannot be eliminated, no matter how much the carburetor hold-down fasteners are tightened. The best way to eliminate this problem is to replace the gasket when the propane carburetor is installed. However, if you must re-use one of these gaskets, it is often possible to remove the bushings and grind the ends to shorten the length of the bushing slightly. This will provide some additional gasket crush before the carb or mixer seats on the bushings.

If there's a final troubleshooting bit of advice we can give, it's to *think things through completely before you start, work in a safe and professional manner, and don't be hasty to pass judgement on a new propane installation*! In a gasoline engine, problems are routinely handled without being blamed on the type of fuel, and the propane system is much less complicated than gasoline. When you are "fiddling" with the propane components keep sparks and open flames away from the work area! Whenever you fill a propane tank you'll notice that the storage tanks are always marked "Danger-No Smoking Or Open Flame Within 50 Feet." This advice applies to work areas as well! And remember, don't release propane vapors inside a garage where there is either a water heater with a pilot light or a gas or oil-fired space heater. We don't want to lose any readers, and any mishap—even one caused by ignorance or carelessness—reflects badly on a fuel that is safe, efficient and clean-burning.

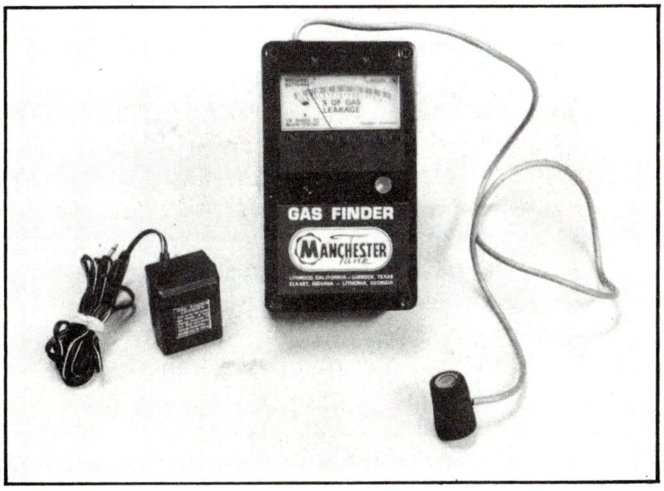

Manchester Tank markets a tool that is invaluable for troubleshooting leaks in the system—an extremely sensitive gas detector that is cordless and comes with a charger. Every professional installer should have a reliable tester for initial safety checks.

SOURCES

SCHOOLS

If your interest in propane performance runs deeper than the conversion of your own car, you might consider one of the industry's technical schools on propane carburetion. Perhaps you'd like to get into this growing field in a professional way; but even if you just have a tremendous thirst for new knowledge, these intensive technical seminars may be just what you're looking for.

For many years, Impco Carburetion, in Cerritos, CA, has been a leader in the gaseous-fuel carburetion field, manufacturing products that are in use all over the world. After 30 years of contact with O.E.M. and aftermarket customers (to whom Impco has always provided valuable after-sales technical assistance), they realized the importance of good tech information toward using gaseous fuels successfully. To this end, several years ago, Impco initiated an in-house carburetion school. Designed to train mechanics who maintain fleet vehicles using Impco equipment, the three-day intensive course is packed with enough solid information to be of value even to those who have been in the gaseous-fuel carburetion industry for years.

Not everyone has had a college education in physics or chemistry, so the course is taught on a level from which everyone can benefit. It begins with a basic discussion of air and gas properties and wastes no time in getting down to serious discussion on how to make an engine respond best to gaseous fuels, how to take advantage of gas properties, and how to avoid common mistakes. Engine building, valve and head work, and ignition requirements are stressed throughout the course—with most of the information applying to any maker's carburetion equipment, not just Impco's. From classroom work, blackboard illustrations, and actual cutaways, the class progresses in its third day to Impco's research laboratory, where the student is given a rare opportunity to view the inside of a running engine. The lab is equipped with a single-cylinder research engine built with a thick *glass* cylinder wall, so the student can directly observe the effects of various carburetion and ignition adjustments in the combustion chamber. This learning is further backed up with study of a real six-cylinder Ford engine running under load on an engine dynamometer hooked to a scope, exhaust-gas analyzer, and spark-plug thermocouple that directly reads combustion-chamber temperatures.

Impco's three-day course is generally given on Tuesday, Wednesday, and Thursday of the third week in every month. The course is free to those who qualify, and attendance is usually overflowing with students from all over the U.S. and some foreign countries. So be advised that the course is usually booked months in advance. For more information about the carburetion course, contact Impco Carburetion Inc., 16916 Gridley Place, Cerritos, CA 90701. Or call them at 213-860-6666.

In case you're not conveniently located near Impco, propane carburetion classes are also offered by the following companies:

Irving Krutsinger
The Dancey Co., Inc.
P.O. Box 16188
Tampa FL 33617
813-626-2102

Don Schultz
Schultz Gas Service
2801 East 175 Street
Lansing IL 60438
312-474-6555

Jack Vranken
Propane Equipment Corp.
P.O. Box 236
Shrewsbury NJ 07701
201-747-3795

A program called "The Annual Liquified Petroleum Gas Technical Service School" is offered by the Agricultural Extension Service of the University of Minnesota. This is an intensive three-day seminar, usually conducted in March of each year. The program covers all aspects of LP service, including: piping and transmission, bulk handling, domestic heating, burner service, customer relations and LP gas carburetion. (As testimony to the long history of LPG service, the current program is the 38th annual seminar.) The seminar costs $50 per attendee, and a certificate is given for completion of all three days of the program. Enrollment is on a "first-come, first-serve" basis. Inquiries should be directed to: University of Minnesota, Office of Special Programs, 405 Coffey Hall, 1420 Eckles Ave., St. Paul, MN 55108, or telephone (612) 373-0725.

Several programs on both general LP service and automotive LP carburetion are available at the North Dakota State School of Science. Basic and advanced workshops are offered in all aspects of LP carburetion, including: emissions controls, troubleshooting and engine conversions. This school offers courses for individual students or they will also conduct "company schools," in which they address the specific needs and questions of the participating company's fleet mechanics. For more information contact: Everett Holtz, North Dakota State School of Science, North 6th St., Wahpeton, ND 58075, or call (701) 671-2101.

CORRESPONDENCE COURSES

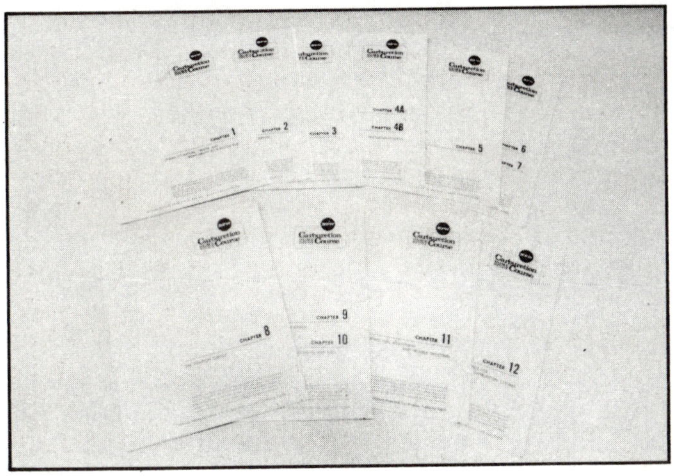

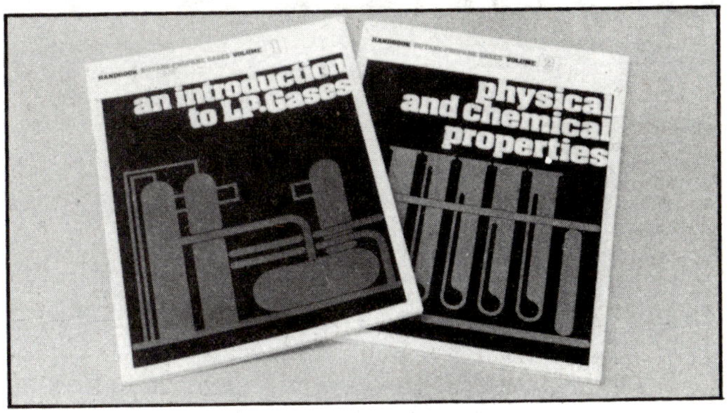

In addition to all those subjects you've seen advertised on match folders, home-study correspondence courses on propane carburetion are available. The offices of Butane-Propane News (338 East Foothill Blvd., Arcadia, CA 91006, phone 213-357-2168) offer a 12-part carburetion course with 102 pages of material. For your $25 fee, they send you two chapters at a time, each with tests at the end. When you send back your tests, they grade them and send you the next two chapters.

Beginning with an apt chapter on "LP-Gas, Its Nature, Origin, and Adaptability As a Motor Fuel," the course continues with chapters such as "Principles of Engine Operation," Operating Temperatures, and "The Ignition Circuit." Much of the material is devoted to explaining the principles under which an internal-combustion, spark-ignition engine operates; but the final four chapters deal with "Pre-Conversion Checks," "How To Make Adjustments and Tests," "Starting-Fuel Adjustment And Troubleshooting," and "Preventive Maintenance for LP-Gas Carburetion Systems."

In addition to their course, Butane-Propane News has several other books of interest, including a two-volume *Handbook on Butane-Propane Gasses.* Volume One is titled *An Introduction to LP Gasses*; its 47 pages deal with the makeup and production of the gasses along with illustrations of the many uses of LPG fuel today ($3 per copy). Volume Two deals extensively with the chemical and physical properties of LP gasses and is more for the scientifically-oriented readers ($5 per copy).

Another correspondence course in propane carburetion, offered by Krebbs Associates Inc., is aimed mainly at companies who want in-house control of training their fleet mechanics. Written by the late Jack Krebbs (who also wrote the text for the Butane-Propane News course), it consists of an instructor's manual that comes with slides, cassette tapes, and four student manuals. This course deals as much with after-conversion maintenance as it does with installation and has been used successfully for both small and large fleets, such as Midas and Dancey Company. The instructor's manual, slides, cassettes, and four student manuals cost $245; additional student manuals are available for $22 each—after the basic course has been purchased. For more information, contact Krebbs Associates Inc., 1137 North Old Manor Road, Wichita, KS 67208, or phone 316-683-5257.

A reminder of how long propane has been around as a widely-used fuel is the story of Butane-Propane News. Begun in 1931 as an editorial news section of a local—and later national—magazine called *Western Gas*, it became a monthly publication of its own in 1939.

Aimed primarily at the distributor-dealer market, the 60-page magazine deals with subjects that most concern people in the propane industry and the energy business in general: government regulations, safety, future prospects and trends—as well as some information valuable to the professional propane installer.

Things move fast in the dynamic energy business; so to keep readers informed of late-breaking news, Butane-Propane News also publishes two newsletters: "The Weekly BPN Propane Newsletter" deals with much the same material as their magazine but with up-to-date market statistics and other timely features; the "International Butane-Propane Newsletter" deals mainly with the larger scope of worldwide developments in the industry (prices, plants, finance, production, demand levels, and world movements). These publications can be obtained from Butane-Propane News Inc., 338 East Foothill Blvd., Arcadia, CA 91006. Phone 213-357-2168.

SOURCES

REFILLING DIRECTORY

If you're driving a diesel-powered vehicle and do any traveling at all, you can't be without the diesel source book published by Mercedes-Benz; now, an equally-helpful publication is available for propane sources. Published by the National LP-Gas Association (NLPGA), the book lists some 6600 locations where you can tank up your RV or camper with propane for the stove and 3780 dealers who sell propane for motor fuel.

The 1983-84 edition (a new edition is due sometime in 1986) of the *LP-Gas Refilling Directory* contains 25% more listings than the previous edition and 9 pages of special safety-related tips on propane systems for the consumer.

This is an extremely handy publication to have on hand when you're making the vacation trek into unfamiliar territory. Of course, not every possible source for propane is listed here; but those that are make trip-planning a little easier. To order, send $3.50 per copy to *1984 Refilling Directory*, NLPGA, 1301 West 22 Street, Oak Brook, IL 60521. Ask them to keep you on their mailing list for notice of revised editions.

Besides the extensive listings in the *NLPGA Refilling Directory*, the traveler's best listing of up-to-date propane sources is your local telephone Yellow Pages. Additionally, many of the larger propane distributors around the country publish booklets that list their own dealers around the country. Some of these companies are:

Empire Gas Corp.
PO Box 303
Lebanon MO 65536

Getty Refining & Mktg. Co.
PO Box 1650
Tulsa OK 74102

Northern Propane Gas Co.
PO Box 475
Columbus NE 68601

Pargas Inc.
PO Box 67
Waldorf MD 20601

Petrolane Inc.
PO Box 1410
Long Beach CA 90801

Pyrofax Gas Corp.
PO Box 2521
Houston TX 77001

Suburban Propane Gas Corp.
PO Box 206
Whippany NJ 07981

Williams Energy Co.
PO Box 3478
Tulsa OK 74101

PROPANE CARBURETION EQUIPMENT

MANUFACTURERS

American Super Sprint Systems, Ltd.
Box 433
Pleasantville NY 10570
(importers of "Bedini" dual-fuel carburetion)

Beam Products Mfg. Co.
3040 Rosslyn St.
Los Angeles CA 90065

Century LP Gas Carburetor
Marvel-Schebler Div. of Borg-Warner Corp.
707 S. Side Drive
Decatur IL 62525

Impco Carburetion Inc.
16916 Gridley Place
Cerritos CA 90701

Landi-Hartog USA
3208 S. Hutchison Ave.
Los Angeles CA 90034

OHG Inc.
10065-A S. Greenleaf
Santa Fe Springs CA 90670

Procal Corp.
1151 Dove
Newport Beach CA 92660
(distributors of Dutch "Vialle" dual-fuel carburetion)

OTHER LPG-RELATED EQUIPMENT

Ak Miller Enterprises **9238 Bermudez** **Pico Rivera CA 90660**	Distributors of OHG carburetion equipment and AiResearch turbochargers. They offer complete turbo-propane kits and have many years of experience in carburetion and turbocharging.
Almac Engineering Inc. **2079 Mound Rd.** **Warren MI 48091**	Manufacturers of remote-fill receptacles with locking covers.
Autotronic Controls Co. **6908 Commerce** **El Paso TX 79915**	Manufacturers of the MSD line of ignition wires, Dual-Curve ignition modules, and other ignition products.
Brown Manufacturing **PO Box 844** **Bedford IN 47421**	Manufacturers of automatic stop-fill valves for LPG tanks.
Cann Products **9540 Stewart & Gray Rd.** **Downey CA 90241**	Manufacturers of advance-limit plates, water adapter Y's, and H's, and many other installation aids.
J&S Electronics **13925 Parkway Dr.** **Garden Grove CA 92643**	Manufacturers of Injectamatic electronic water injection.

MANUFACTURERS OF PROPANE MOTOR-FUEL TANKs

American Welding & Tank Co.
PO Drawer JJ
Plant City FL 33566

Brunner Engineering & Mfg. Inc.
PO Box 1367
Bedford IN 47421

Cal-Tank
PO Box 4427
Redding CA 96099

Energy International
PO Box 1346
Elkhart IN 46515

Roy E. Hanson Jr.
1924 Compton Ave.
Los Angeles CA 90011

Isometrics Inc.
PO Box 660
Reidsville NC 27320

Manchester Tank Co.
2880 Norton Ave.
Lynwood CA 92062

Wendland Mfg. Co.
PO Box 808
San Angelo TX 76902

GLOSSARY

20% valve See outage valve.

air-valve cup A passageway and seat for the gas jet of a propane mixer. During engine operation, manifold vacuum pulls the gas jet away from the air-valve cup, allowing propane to flow through the mixer and into the intake manifold. If the air-valve cup and jet do not operate properly, excess propane may enter the engine (especially at idle and low speed), resulting in erratic idling and a loss of fuel economy.

alkyd An excellent electrical insulating material used to manufacture specialized ignition distributor caps, rotors, and coil components. Alkyd is usually very distinctive in color (tan, blue, red, etc.) and easily identifiable.

alternative fuel A fuel that takes the place of a typical or standard fuel. Propane and natural gas are alternative fuels for conventional gasoline-fueled automotive engines. Alternative fuels offer some advantages (clean burning, efficiency) and some disadvantages (refilling stations are less abundant, converting an engine to burn an alternative fuel requires additional expenditures for special fuel components, etc.).

boost controller A device used in a turbocharging system to limit the boost pressure delivered to the engine. A boost controller is a spring-controlled valve installed in the air tract between the turbocharger and the intake manifold. When boost reaches a prescribed limit, the valve begins to close, restricting intake flow into the engine. A boost controller, however, does not limit the pressure delivered by turbocharger, rather it creates backpressure that reduces the efficiency of the turbocharger. Boost controller valves are less expensive than wastegate limiters.

compression ratio The ratio of the maximum cylinder volume (when the piston is at bottom dead-center) to the minimum cylinder volume (when the piston is at top dead-center). The most accurate and straightforward method of determining compression ratio is to perform accurate measurements during engine assembly. Move a piston to BDC and seal the piston to the bore with light grease. In a similar fashion, seal the corresponding intake and exhaust valves for that cylinder to the valve seats, and attach the cylinder head (with a gasket) to the block. Fill the cylinder through the spark-plug hole with a carefully measured liquid from a burette or similar graduated measure. (To prevent bubbles from forming in the cylinder, the engine should be rotated so the spark-plug hole is at the top). Then drain the liquid, move the piston to TDC, and repeat the process. The larger volume (at BDC) divided by the smaller volume (at TDC) is the compression ratio.

converter/regulator Converts liquified gaseous fuel (pressurized) to a gaseous state by allowing it to expand. The converter (sometimes called a vaporizer) is commonly used on propane- or LNG-fueled engines. It is a simple expansion chamber, however in automotive applications, there is another separate chamber in the converter through which warm engine coolant is circulated. This prevents the converter from freezing as the liquid turns to vapor. The converter often incorporates a pressure regulator that reduces the vapor to an appropriate pressure before it is delivered to the mixer (a gaseous fuel carburetor).

detonation The secondary ignition of the air-fuel mix in the combustion space, resulting in excessively rapid burning and extreme pressures. Detonation is caused by inadequate fuel octane, high temperatures, improper combustion chamber design, excessively lean air-fuel ratios, etc. Detonation produces dangerously high stress on the cylinder block and the operating components of the crank assembly. In severe or excessive cases this can lead to engine failure. Detonation is different from pre-ignition, in that detonation requires two simultaneous combustion fronts (fuel burning in two or more places in the combustion chamber); whereas pre-ignition occurs when the fuel-air mix ignites in a localized portion of the chamber before the spark plug initiates combustion. Both pre-ignition and detonation produce an audible "knock" or "ping," but detonation does not produce the rapid "wild pinging" noise that is associated with pre-ignition.

draw-through A supercharging system that draws a fuel-air mix from a carburetor or injector and forces the pressurized mixture to the engine. A draw-through supercharging system compresses both air and vaporized fuel; while a blow-through system compresses only the inducted air.

dual-fuel conversion Installing a second fuel system—usually gaseous fuel—on a vehicle to supplement (or relegate to a secondary source) the standard fuel system. Switching from one fuel to the other is often accomplished by a dash-operated control that turns on/off the appropriate fuel source and/or pumps, directs air into the carburetor or mixer, and (optionally) switches between two ignition advance curves.

dynamic compression ratio In a supercharged engine, during boost, the dynamic compression ratio is higher than the static (or measured) compression ratio because the fuel-air mix inducted to the cylinders is more dense than the comparable mixture inducted under non-boost conditions. In other words, a greater quantity of fuel-air is inducted to the chamber of a boosted engine, and though the mechanical compression remains unchanged, the effective compression ratio will be greater because a greater quantity of air-fuel is being compressed into the minimum volume (at TDC). Because of this phenomenon, the static (mechanical) compression ratio on a supercharged engine must be lower that a comparable non-supercharged engine (all other factors being equal) to prevent detonation difficulties. Conversely, in a normally aspirated engine, the density of the air-fuel mix drawn into the engine may be reduced due to restrictions and inefficiencies in the induction system. In this case, the dynamic compression ratio is less than the static ratio. However, this is rarely the case in a high-performance engine at full throttle because the intake system is generally very efficient; sometimes to the point of slightly increasing the induction density due to ram effects in the port runners.

excess-flow valve A safety valve used in LPG and LNG fuel systems to completely shut off fuel flow if the rate of flow exceeds a preset maximum. A standard part of all gaseous fuel automotive installations, the excess-flow valve will stem fuel flow when a pressure line ruptures or for some reason an unrestricted flow of fuel occurs. This is a safety feature to prevent fire or explosion. The excess-flow valve is installed at the tank outlet.

fill valve A valve on liquified gaseous fuel tanks (LPG and LNG) that is used to refill the tank. When access to the tank is limited, the fill valve is sometimes mounted in a remote location. Also see outage valve.

filter/fuel-lock or **filter/shut-off** A device used on liquified gaseous fuel installations that filters particulate impurities from the fuel and shuts off the supply of fuel when the engine is not running. The filter/fuel-lock performs an important safety function: it prevents fuel from collecting in (and leaking out of) the induction system when the engine is not running. This device is designed to handle gaseous fuels while they are still in a liquid form.

gas jet A component in gaseous fuel carburetors that controls the flow of fuel into the inducted airstream. The gas jet assembly is connected to a vacuum diaphragm that reacts to induction airflow; as flow increases, the gas jet is pulled further from the seat (also see air-valve cup) and more gaseous fuel enters the inducted airstream. The shape of the gas jet determines the fuel delivery at a prescribed vacuum level.

grit A systemized rating of surface texture used to categorize sandpaper, grinding stones, honing stones, etc. Surfaces rated at 60 grit or less are very coarse, from 60 to 120 grit are coarse, from 120 to 200 grit are medium; from 200 to 400 grit are fine; from 400 to 600 are very fine; and above 600 is considered a polishing surface.

hydrostatic-relief valve A relief valve that vents potentially dangerous excessive pressure in the feed line from an LNG or LPG (propane) tank. In automotive gaseous fuel applications, if the feed line from the fuel tank is heated (perhaps by a nearby portion of the heat-soaked exhaust system) when the engine is shut off and the tank valve is closed, the line pressure can become dangerously high due to vapor expansion. In these situations, the hydrostatic-relief valve will vent the excess pressure to the outside of the vehicle. This reduces the possibility of damage to the line, valves, or the regulator/vaporizer.

ignition curve The change in ignition advance (caused by the centrifugal advance system) relative to engine speed (rpm). When the relationship between ignition timing and engine speed is plotted on a two-axis graph, the result is a distinctive curve (although at times the "curve" may appear to be a straight line). In general, as engine speed increases, the amount of ignition advance must also increase. Though limited by other phenomenon, the increase in timing allows more time for the combustion process to occur before the power stroke concludes. However, the precise advance at all engine speeds can only be determined by specific engine and application requirements.

inches of mercury and **inches of water** A standard for measuring pressure differentials, as compared to standard atmospheric or ambient pressure. In automotive applications a pressure differential less than standard atmospheric is usually called a "vacuum" and a differential greater than atmospheric is usually called "boost." Typically, pressure differentials are measured with a manometer (also see manometer). A one pound per square inch differential (from atmospheric pressure) will displace a column of mercury 2.04 inches in a manometer. One psi of pressure differential will displace a column of water 27.72 inches.

induction hardening A process that rapidly heats and subsequently cools (at a specific rate) a metal component to increase the hardness of the material. The metal component is usually positioned inside a heavy wire coil, through which a large electric current flows. A powerful magnetic field is produced, which raises the metal to a red heat in just a few seconds. The component is then cooled by a controlled liquid or air quench. This process will change the microstructure of the metal, increasing the strength and adding a surface or deep hardness to the material.

interference angle An intentional mismatch of the valve face and valve seat angles. Grinding the angle on the valve face about one degree less (usually 44 degrees) than on the valve seat (normally cut at 45 degrees) is a traditional practice of engine builders. Called an "interference-angle valve job," it is said to help initial sealing during the first few hundred miles of engine operation. However, some engine builders believe that interference angles are not desirable on a precision valve job. According to this line of thinking, if modern valve facing tools are correctly and carefully employed, valve seal will not be a problem. And in fact, an interference angle can cause the exhaust valve to overheat. Since there is a reduction in the contact area between the seat and the face, the heat transfer from the valve head to the seat will be reduced. This is of particular importance in a race engine or when the combustion temperatures are unusually high (as with propane).

latent heat of vaporization The heat required to change a substance from a liquid to a vapor (to boil). A substance in vapor form (gaseous) has a higher energy level than that same substance in a liquid state. When heat is applied to a liquid (such as water), its temperature will rise until it reaches boiling temperature, then it will continue to absorb heat (draw in heat energy) without an increase in temperature until boiling occurs; this increase in energy without an increase in temperature is the heat of vaporization. When a liquid fuel (such as gasoline, alcohol, etc.) vaporizes (evaporates), the heat of vaporization absorbed by the fuel draws in heat, resulting in a cooling of the air-fuel mix and the induction system. In an LPG or LNG equipped engine, the heat of vaporization is so high (for liquified gaseous fuels), that a special device—the converter or vaporizer/regulator—is used as a "boiling pot."

leak-down test An integrity test of the seal provided by the rings, valve seats and head gaskets. A leak-down tester pressurizes a cylinder (with compressed air) and measures the amount of escaping gas. The resultant "leakage" is displayed on a gauge as a percentage. A 10% leakage is usually considered "normal," while many professionals, using precision engine components, machining and assembly procedures, can obtain readings less than 5%. When the leak-down percentage is lower, the engine produces a tighter mechanical seal and will operate more efficiently.

LNG or **liquified natural gas** Natural gas stored in a liquid form. Natural gas can be maintained in a liquid form in high pressure tanks (LNG must be compressed to over 2000psi to maintain a liquid state at atmospheric temperature) or in special super-cold storage tanks; LNG boils at -258 degrees F (-161 degrees C).

LPG or **liquified petroleum gas** Propane gas stored in liquid form. Propane gas can be maintained in a liquid form in high pressure tanks (LPG must be compressed to about 180psi to maintain a liquid state at atmospheric temperature) or in special cold-storage tanks; LPG boils at -43.7 degrees F (-42 degrees C).

manometer A simple device for measuring relative pressure. A manometer is a U-shaped tube (glass or clear plastic), containing liquid, with one end open to atmospheric pressure and the other end connected to the test pressure. When mercury is used in the manometer tube, a differential (from atmospheric pressure) of one pound per square inch (psi) will displace the mercury column 2.04 inches. However, when water is used in a manometer, the sensitivity is substantially increased: a differential of one psi will displace the water column 27.72 inches.

mixer A carburetor for gaseous fuel. Most mixers operate on a constant pressure-drop principle, in which a diaphragm controls the gas jet (a gaseous-fuel equivalent to the main jet and metering rod of a conventional carburetor). When manifold vacuum reaches about 7 inches of water (below atmospheric), the diaphragm opens the gas jet orifice and allows fuel to flow into the engine. As manifold flow increases, the gas jet is opened farther, permitting additional fuel flow. This principle is used to meter fuel flow and to prevent gas from flowing out of the carburetor when there is no manifold vacuum present (e.g., when the engine stalls).

natural gas A flammable gas found within the crust of the Earth with a boiling point of -258 degrees F (-161 degrees C) at atmospheric pressure; a very important petroleum fuel. Natural gas consists primarily (85 to 95%) of the hydrocarbon methane, but proportions of other compounds are also present (e.g., 10-18% ethane, 0.7-2% propane, 0.2-0.7% butane, and other light hydrocarbons). Natural gas is piped to millions of homes for heating, is used widely in industrial and commercial applications, and is used as a motor fuel.

octane A relative rating system that expresses the resistance of a fuel to detonation (knock). Octane is measured in several standard tests, each of which produces a unique octane number (research octane, motor octane using benzene, or motor octane using ethyl alcohol, etc.). Most "fuel pump ratings" are an average of research octane and motor octane ratings. The antiknock properties of gasoline can be improved by adding chemical agents, the best of which is tetraethyl lead. However, public concern about the lead content of automotive emissions has led to federal legislation requiring the almost exclusive use of unleaded gasoline in recent production vehicles. The elimination of tetraethyl lead is the primary reason for the reduction of pump octane ratings over the past few years.

outage valve A small manually-operated valve on compressed liquified-gas tanks that alerts filling personnel when the tank has reached maximum capacity. When the outage valve is opened (and the tank is not full) a slight amount of fuel vapor will escape from the valve. When the tank is full, liquid will flow from the valve. Note: All LPG and LNG tanks are considered full when liquid escapes from the outage valve, however, the tanks are not actually full. A "full" tank will retain an inch or two of expansion space, essential to prevent excessive pressure when the tank is heated.

pressure drop The pressure (usually of a moving air mass) measured in one place/area that is lower than the pressure measured in another place/area; e.g., the pressure drop (relative to ambient pressure) in the cylinder as the piston moves down on the intake cycle; this reduction in pressure draws air into the carburetor, manifold, and cylinder.

pressure-regulator valve A valve—working against a spring—designed to allow the flow of a gas or liquid only when a preset pressure is reached. Pressure regulator valves are used to maintain a constant pressure, and their operation is considered a normal part of system function; e.g., oil pressure regulator valves in engines and transmissions, etc. Also see pressure-relief valves.

pressure-relief valve A spring-loaded "pop-off" valve that is designed to allow the rapid flow of a gas or liquid only when a preset pressure is reached. Pressure-relief valves are used to relieve potentially dangerous pressure levels and their operation is not a normal part of system function; e.g., air-compressor or propane tank blow-off (excess pressure) valves. Also see pressure-regulator valves.

propane A hydrocarbon fuel with a molecular weight of 44.09 and boiling point of -42.1 degrees C (-43.8 degrees F) at atmospheric pressure; a major constituent of LPG. The net heat of combustion for propane is 21,646 Btu/lb; the stoichiometric air-fuel ratio is 23.8:1. Propane is a practical fuel for motor vehicles because it is easily stored, produces good performance, and can increase engine life (compared to gasoline). Equipment to install propane systems is readily available, and propane is widely available and reasonably priced.

relative pressure A method of measuring pressure whereby atmospheric or ambient pressure is considered zero on the scale. A complete vacuum is -14.7psi relative.

research rating See octane.

road tax A tax imposed on those that use the roads and highways, usually levied on motor fuel, tires, etc.

rough honing The first stages in the honing process, wherein the cylinderwall surface is honed with a coarse-grit honing stone. See also grit.

single-plane manifold An intake manifold that places all the port runners on the same level and—usually—connected to a common plenum area (an open-plenum design); in this design each cylinder draws fuel-air mix directly from the entire plenum and the entire throttle bore area of the carburetor. A single-plane manifold often provides an increase in peak horsepower, although it can, depending on the manifold design, be accompanied by a reduction in low-speed torque and throttle response. The short, direct runners (typically) can reduce flow velocity and turbulence while improving fuel distribution—adding to high-rpm flow and performance—but the same reduction in flow velocity can reduce fuel atomization and low-speed efficiency. This is not a problem with gaseous fuels, however, and many engines fueled with propane or LNG provide excellent overall performance with a single-plane, open-plenum manifold. Most single-plane manifolds are also open-plenum manifolds (i.e., the plenum is a single chamber), however some racers have found that adding a partial plenum divider can improve mid-range horsepower without substantially reducing maximum power.

spark-plug insulator See center insulator.

static compression ratio The measured compression ratio, not taking into account any dynamic effects (e.g, supercharging), ram-type intake manifolds, etc. See compression ratio.

stoichiometric air-fuel ratio The chemically correct proportion of air to fuel—by weight—that is required to theoretically oxidize (burn) the fuel completely. When just enough air is added to the fuel so that after combustion (ideal) all of the fuel is consumed and there are no remaining oxygen molecules, the air-fuel ratio is considered stoichiometric. Generally, the stoichiometric ratio (for a particular fuel) is considered to be the leanest air-fuel ratio for efficient combustion. In practice, however, a richer mixture is usually required to obtain optimum performance and driveability. Typical stoichiometric ratios are: for gasoline, 14.95:1; for diesel oil, 15.06:1; for propane, 23.87:1, for natural gas, 17.24:1; and for ethyl alcohol, 9.01:1. See also Btu.

valve recession A phenomenon wherein the exhaust valve erodes the valve seat and the valve "recesses" into the cylinder head. Valve recession is caused by excessive friction or heat at the contact surface between the valve face and valve seat. Excessive friction is often due to the use of unleaded fuels that do not offer the lubricating properties of tetraethyl lead additive. Excessive heat can occur during prolonged detonation or when certain fuels are used that burn hotter than gasoline (such as propane). Many late-model engines utilize hardened valve seats that function well with higher friction and heat levels. The hard surface of these seats have a better coefficient of friction with the valve face and can eliminate recession. Hardened valve seats are produced by induction hardening the stock seats or by replacing them with hard-seat inserts.

valve-seat insert or **hard-seat insert** A hardened steel replacement for the standard cast-iron portion of the cylinder head that contacts the valve face (at the terminus of the port and combustion chamber). Properly installed inserts (a tight press fit in a precision-machined cavity) will withstand many valve jobs, although they are difficult to grind. Hard-seat inserts will not transfer heat from the exhaust valve quite as efficiently as the stock seat, since the heat must pass through the insert and subsequently through the interface between the insert and the head. The quality of the machinework performed by the hard-seat installer has an important effect on the long-term reliability of this modification (a loose press fit can allow the seats to back out and result in expensive failure).

wastegate A boost-actuated valve on a turbocharged engine that prevents induction pressures from prevents detonation or stress damage to the engine and turbocharger. The wastegate senses the boost pressure in the induction system and when a preset level is attained (this level is adjustable, sometimes with a remote control in the driver's compartment), the valve (gate) opens and permits a portion of the exhaust gases to bypass the exhaust-driven element of the turbo. This effectively reduces the output pressure from the turbo compressor.

"BOOK USER'S GROUP"

Send in your name and address now to become a member of the S-A Design Book User's Group (BUG)—its free! As a member you will receive periodic newsletters containing reader feedback, interviews with authors, technical information, updates on new books, and much more. Plus, all BUG members qualify for special discounts. Just mail in the attached membership form (or a copy) to:

S-A Design Books
Book User's Group
515 West Lambert, Bldg. E
Brea, CA 92621-3991

BOOK USER'S GROUP—BUG

Name ______________________

Street ______________________

City ______________________

State ______________ Zip ______

Do you need a catalog of S-A Design books?

YES☐ NO☐

BOOK USER'S GROUP—BUG

THE CHEVROLET RACING ENGINE is the "bible" of racing engine preparation. Written by renowned engine expert Bill "Grumpy" Jenkins and crammed with tech tips for all racing enthusiasts. 160 pages, over 300 illustrations.

SUPERPOWER provides practical bolt-on info on today's hottest performance topics. Full descriptions, evaluations, installation tips and source lists of kits and suppliers. A must-have for your library! 160 pages, over 320 photos.

MOPAR PERFORMANCE is the best single source for performance info on smallblock and bigblock Chrysler engines. Includes: head and block prep, oiling system mods, factory tricks, airflow evaluations, and much more. 112 pages, over 200 photos.

THE COMPLETE GUIDE TO BOLT-ON PERFORMANCE is a tremendous money-saver for the do-it-yourself performance mechanic. Perf. equipment shown with step-by-step photos, buyers guides. 160 pages, over 600 photos.

FORD PERFORMANCE is the best book available on Ford engine science, period! Loaded with smallblock, bigblock and Cleveland info, this guide includes enerything from the basics to tech tips, and racing secrets. 128 pages, over 400 photos.

CHEVY PERFORMANCE is an in-depth source for the latest technology on blocks and heads, ignition, carburetion, manifolding, cams and valvetrains, ring and borewall prep, and more. It's all here in 128 pages, over 400 photos and many charts.

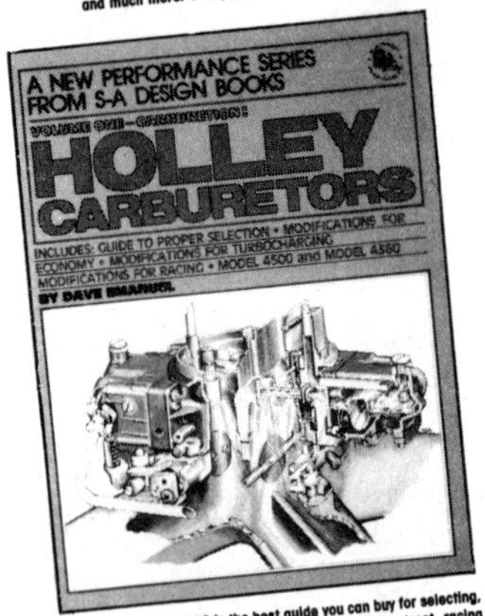

HOLLEY CARBURETORS is the best guide you can buy for selecting, tuning and modifying all popular Holleys. Mods for street, racing economy and turbocharging included, plus much more. 128 pages, over 350 photos.

PERFORMANCE WITH ECONOMY will show you how to get great performance with good gas mileage. Impossible? No! Covers low-restriction mufflers, turbos, water and nitrous injection, much more. A great engine builder's reference. 128 pages, over 300 photos.

CUSTOM PAINTING will show you how to produce eye-poping paint tricks in your own driveway. Includes: surface prep, choosing paint, repainting/touchup, flame painting, murals, and much more! 128 pages, 500 photos.

V-6 PERFORMANCE is a compete look at the new generation of lightweight V-6 engines. Includes Chevy 60/90 deg., GM/Buick and Ford 60 deg. engines. Info from experts, including Smokey Yunick, Bill Jenkins. 128 pages, over 350 photos.

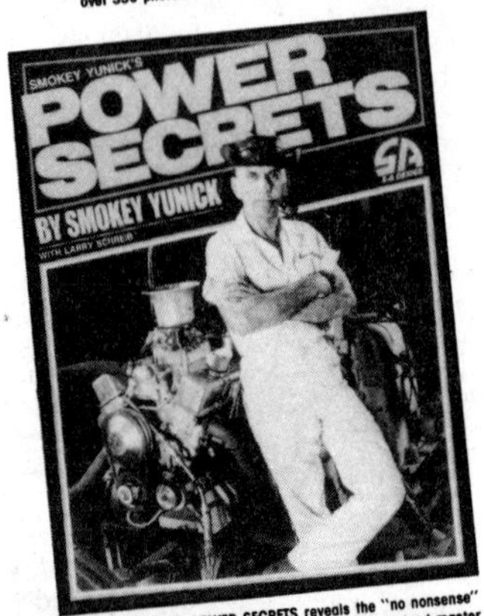

SMOKEY YUNICK'S POWER SECRETS reveals the "no nonsense" tricks that have made Smokey Yunick the acknowledged master of engine development. Filled with mechanical insights that every performance enthusiast and racer will enjoy. 128 pages, over 200 photos.

STREET SUPERCHARGING will get you off and running...fast! Find out which blowers work and why, where to get the parts, and how to install them. Get into supercharging the right way, get STREET SUPERCHARGING. 128 pages, over 350 photos.

We encourage individuals to purchase books from their local retailer. S-A Design Books are sold internationally in speed shops, book stores, and automotive parts outlets. However, if you cannot find our books locally, you may order direct from S-A Design by pre-paying $12.95 plus $2.00 postage and handling per copy (California residents add $.84 tax each). Mail to: S-A Design Order Desk, 515 West Lambert, Bldg. E, Brea, CA 92621-3991.